Industrial Vacuum 101

Part One: The Basics of Vacuum Technology

SECOND EDITION

Dan Bott

Industrial Vacuum 101

Part One: The Basics of Vacuum Technology

SECOND EDITION

Dan Bott

Table of Contents

Table of Contents – Continued

Acknowledgements

Thanks to all of you who have asked those difficult questions over the years. Through answering these questions, it has been my great fortune to study, learn about and work with vacuum applications. They have expanded my own learning and knowledge of vacuum processes and the many industries where vacuum is used. I hope I have done justice with the answers, as my primary goal is to serve my customers. Also, special thanks to the many editors who have helped me through the process of writing a book and for the many hours they have spent reviewing, changing and advising.

Introduction

Welcome! This book will be your guide on the science behind vacuum technology and vacuum applications. I have had extensive training and experience working with manufacturers of both vacuum pumps and accessories to vacuum pumps. In addition, as a vacuum systems consultant, I have worked on many specialized projects where the primary goal was to optimize a vacuum system in a manufacturing or service application. Each year, users of vacuum equipment waste millions of dollars due to misinformation regarding the use of vacuum pumps, vacuum piping and vacuum accessories. Misapplied vacuum equipment not only results in increased capital costs, but also in increased operational and energy costs. More directly to the manufacturers of goods and services, poor production efficiency can have its roots in poor performance from the process vacuum system. While the costs associated with production down time or "slow time" can be difficult to quantify, they are every bit as important as other bottom line considerations.

This technical series is designed with one objective: a direct, straightforward instruction on basic vacuum technology that will provide the new user with the necessary tools to communicate effectively with manufactures and users of vacuum equipment. This series is also designed to help users make informed decisions regarding the use of vacuum equipment. Part one (this book) assumes no prior knowledge of vacuum equipment although some exposure to the world of pneumatics is helpful because of the familiarity with fundamental industry terms and concepts. Basic math skills are necessary to work problems in the text. The bottom line is that this text is intended for the nonscientist.

How do we begin this journey? Similar to on-site vacuum training programs, we begin with a macro view and work our way down to a micro view. Actually, we should start with a discussion on the program itself. This book and is part one of a three part program designed to build a better understanding of vacuum technology so that decisions regarding vacuum pumps, vacuum applications and vacuum accessories can be made correctly and with confidence. Part one is essentially the introduction to what is known as 'rough' vacuum technology where you will learn the fundamental aspects of vacuum science. Even the term 'rough' or low vacuum will be explained in detail. Do not panic if you have trouble understanding some of the early concepts and terminology. All will be explained eventually. For example, of the many things you will learn about vacuum science, one will be how air and other gases interact under vacuum based on changes in pressure and temperature. You will also learn the industry buzzwords and terms used every day by vacuum users and process technicians. In other words, you will be able to talk the talk. Most importantly, by studying this book, you will achieve a level of comfort when discussing vacuum pumps and applications that will enable you to communicate effectively and hopefully solve vacuum related problems for yourself or for one of your internal or external customers.

Parts two and three, when complete, will build upon the fundamentals developed in part one and take you further along the learning curve. With these tools, you will be able to tackle some of the more complex application issues seen in today's vacuum system installations. Many seemingly difficult application issues can be solved by using simple rule-of-thumb techniques or variations of standard engineering practices and calculations. Parts two and three assume a completion of part one and a general mathematics acumen. In all cases, terminology and word usage is as user friendly as possible and kept very close to 'street' language. This was done so that an engineering degree is not necessary to achieve a full understanding of the concepts and terminology. Get ready to learn a lot about vacuum technology!

This text incorporates charts, drawings and graphs as necessary to explain each of the highlighted concepts. Just like learning a foreign language, practice makes perfect and that which you do not use tends to slip into the background. At the end of the book in Appendix E is information regarding on-site vacuum technology training and vacuum systems consulting. You can also call me at (251) 609-1429 to schedule an appointment or get an estimate on any vacuum related services.

Also, please send me an email regarding topics you would like added or more fully explained in this book to **dan@dbott.com**. I will be happy to hear your comments and suggestions for improvements.

There is a word of caution to those of you who are new to vacuum technology. Until you are comfortable working with vacuum pumps and related vacuum accessories, it is imperative that you consult with an engineer, consultant or other professional before making significant economic or purchasing decisions based on the information in this book.

Dan Bott
September, 2011

NOTE ON THE SECOND EDITION:
There are several minor changes made to the text and drawings in this edition. The primary goal for the second edition was to make it available again as the first edition copies are out of stock.

Background

Although the average persons' understanding of vacuum science and technology is somewhat limited, vacuum is used in thousands of manufacturing and service applications. Vacuum is used to hold and move aluminum cans through production machinery in can manufacturing plants; it is used for surgical suction in hospitals; it is used as a tool to help paper move faster and more smoothly through printing machinery; it is the medium which moves bulk materials from railroad cars into storage silos; vacuum pumps remove air from packages of meat and other foods so that shelf life is extended; and vacuum is utilized to hold wood, plastic and aluminum sheets in place so that CNC (computerized numerical control) cutting or routing can take place without expensive and time consuming fixturing. These are just a few of the many and varied applications for rough vacuum pumping equipment. Vacuum is used in some form or another to assist in the manufacture of just about every major product in every industrial sector.

Where did the technology of vacuum begin? Actually, the roots of vacuum technology go back to the 1600's in Europe. People like Evangelista Torricelli, Otto Von Guericke and Robert Boyle laid the foundations for the development of vacuum science. To give you one interesting example, in the mid 1600's an enterprising individual named Otto Von Guericke developed a vacuum pump to evacuate a spherical chamber. The chamber was designed in two halves so it could be easily assembled and separated. He took this device and demonstrated the power of vacuum to the royalty of Europe in a unique and effective manner. When Von Guericke put the two hemispheres together and evacuated the air from the inside, the two hemispheres were very difficult to separate. He would use two teams of horses, one on either side of the sphere, to try and pull the hemispheres apart. When the horses failed to pull them apart, he would walk up to the chamber, open a small valve, wait for the air pressures to equalize and then pull the hemispheres apart with his hands. This demonstration astonished and amazed the people of that time! The chambers later became known as the Magdeburg Hemispheres and Von Guericke himself became a celebrity. From these rudimentary beginnings, vacuum science and technology has flourished and advanced to this very day. New technologies, new applications and variations of old technologies are being developed every day enabling manufacturers to more efficiently and effectively produce goods and services.

It is important for you to learn about and understand this unique and fascinating technology because communication about vacuum applications between technicians, engineers, sales/marketing specialists and facilities managers requires a commonality in terms. These terms describe the conditions surrounding vacuum pumps and vacuum applications. Since problem solving in vacuum applications sometimes requires interdisciplinary teams, the understanding of these terms and concepts will enable personnel from these various backgrounds to communicate more effectively with one another. These teams should have the proper foundation of knowledge and experience; this learning series will provide that foundation.

List of Tables and Figures

List of Tables and Figures (continued)

Chapter 1

Pressure Scales

To understand the fundamental basis of vacuum is to understand pressure. Is vacuum really a form of pressure, or is it the absence of pressure? Is it a negative pressure? It is interesting to hear discussions involving vacuum applications and how people describe one level of vacuum or another. Even though most people "know" what vacuum is, it is still amazing to hear the terms used to describe vacuum. The goal in this first chapter is to find a commonality in terms so that we can communicate effectively with one another about each of the various levels of vacuum. When we have a common understanding of the terms regarding pressure and vacuum, we can accurately describe the relative amount of work that can be done with or in a vacuum system. To do an adequate job of describing vacuum, however, it is first necessary to have an understanding of the nature of pressure.

This discussion of pressure starts with a very common term labeled **"atmospheric pressure"** (ATM). Atmospheric pressure at sea level is 14.7 PSIA. Note that most of the discussions in this text will assume sea level elevations and standard conditions. PSIA is a pressure scale that is used to describe many diverse levels of pressure and stands for **P**ounds per **S**quare **I**nch **A**bsolute. What does this really mean? Does it mean that at sea level elevation we feel 14.7 pounds of pressure per square inch pushing down on our bodies? If that were the case, the average human body which has about 3,000 square inches of surface area, would feel a force of 14.7 pounds per square inch times 3,000 square inches. This equates to a total force of 44,100 pounds pushing down on our bodies! That would completely crush all but the toughest of us.

The real answer is that while yes, there is 14.7 pounds per square inch on the outside of each of us pushing in, there is also 14.7 pounds per square inch on the inside of each of us pushing out. The combination of the 14.7 PSIA pushing in versus the 14.7 PSIA pushing out results in a net equalization of pressures. There is no differential pressure between the inside and outside and therefore no uncomfortable crushing of our bodies. This is true not only of our bodies, but also the rooms we sit in, the cars we drive in, the buildings around us and just about every other open container or system we can see or conceive.

The reality about atmospheric pressure is that the earth, which exerts a significant gravitational force on anything that has mass and is relatively close by, holds a blanket of air to its surface. If you took a column of this air that stretched from sea level to the upper end of the stratosphere which was one square inch in cross sectional surface area and weighed it, it would weigh 14.7 pounds. This is where we get the term 14.7 "pounds per square inch". To get the total pressure exerted on the surface of a container or system, it is necessary to add up all the square inches of surface area and multiply that by the ambient atmospheric pressure.

An example would be helpful to further explain this concept. Picture a box or container that has a hinged lid similar in size to a cigar box but much more structurally sound. At ambient conditions when the lid is open there is 14.7 pounds per square inch on the outside of the box pushing in as well as 14.7 pounds per square inch on the inside of the box pushing out. There is a net equalization of pressure or in other words, there is no 'differential pressure'. If the lid of the container was simply closed, does the balance change between the internal and external conditions? No, there is still the same equalized pressure between the two.

Many of you reading this know quite a lot about compressed air applications. What would happen if this closed container was connected to an air compressor and filled with compressed air? The real question is what would happen to the pressure? The answer, of course, is that the pressure would increase. That answers the easy part of the question. The more difficult part of the question is *why* does the pressure increase? To answer this part of the question, it is necessary to look at what happens at the molecular level when air is compressed into a system. Before a quantity of air can be placed inside a closed system and compressed, it must first be captured from the ambient environment. Is compressed air added to a system by scooping up a bucket full of air and pouring it in the top? Of course not! The air must be captured at the inlet of an air compressor by creating a suction effect. The captured air is trapped in a compression chamber, compressed into a smaller volume and only then can it be discharged into a system.

The reason air must be captured is because air molecules are moving very rapidly. They are in rapid motion because they contain relatively large amounts of energy in the form of heat. All air molecules at a temperature above **absolute zero**, which is –460 degrees F, are rapidly moving and colliding with one another. The collisions with one another and the subsequent collisions with the walls of a container are the interactions that create pressure. Stated another way, air pressure is the total force of many molecular collisions per unit area. The original example of 14.7 pounds per square inch is a good description of this principle. It is the force of 14.7 - "pounds", per unit area – "square inches". The more molecules that are in a closed system, the more collisions there are between these molecules within that system. Consequently, there is higher pressure in that system. That is the real reason why the pressure in a closed system goes up when more air is compressed into it.

If the sample container is connected to a vacuum pump instead of an air compressor, it is possible to observe what happens with the pressure under vacuum conditions. As the air is pumped out of the container, the pressure inside begins to decrease. But instead of the pressure being reduced from a higher-than-atmospheric pressure like the air being let out of a tire, the pressure is dropping to levels that are below atmospheric pressure. Keep in mind that there are still molecules inside the system colliding with one another and colliding with the walls of the container, but there are now fewer molecules and fewer collisions. The result of this is lower system pressure.

The more molecules that are taken out of a system, the lower the pressure. One obvious question arises: can every last air molecule be removed from the container, or in other words, can perfect vacuum be attained in a closed system here on earth? For a variety of reasons, the answer is no. First, many of the vacuum pumps used to pull the air out of closed systems are imperfect, that is, at some point, due to design and physical limitations, these vacuum pumps start adding as much air back into the system as they are removing. This is what is known as the **base pressure** of a vacuum system. It is that pressure in a closed system where deeper vacuum (lower pressure) cannot be attained with a particular vacuum pump. Many vacuum pumps are rated at base pressure conditions where the vacuum pump is run with a completely closed inlet to determine the lowest pressure it can attain when not connected to a system. Second, even if there was a perfect vacuum pump, it could still not pull all of the air molecules out of a system. For one thing, air molecules are very small. The size of an air molecule is on the order of 2×10^{-8} centimeters in diameter.

Just a note on scientific notation before we go further. Exponential notation (scientific notation) is used frequently in vacuum work, especially when referring to high vacuum levels. It is a more convenient way to express very large and very small numbers. As an example, 10 to the 3rd (same as 1×10^{3}) is equal to one thousand, 10 squared (10^{2}) is equal to one hundred, 10 to the first (10^{1}) is equal to 10, 10 to the zero power (10^{0}) is equal to 1, 10 to the minus 1 power (10^{-1}) is equal to 0.1, 10 to the minus 2 power (10^{-2}) is equal to 0.01 and so on. Therefore, saying an air molecule has a diameter of 2×10^{-8} centimeters is the same as saying the diameter is 0.00000002 centimeters. As you can see, this is an extremely small particle!

Now back to discussing the small air molecule. Air molecules are so small that they can find their way through many of the seals used on vacuum chambers and interconnecting piping. They can actually travel right through the walls of a container via microscopic fissures and cracks in the materials of construction. In addition to travelling through these small fissures, air molecules can also actually evolve right from the walls a container under high vacuum conditions. When metal is viewed under strong magnification, the surface of the metal does not appear to be smooth and shiny; it actually looks very rough. There are miniature hills, mountains, strands of material and holes scattered throughout.

Air molecules are trapped within these crevices and are released very slowly when exposed to high vacuum conditions. This is a phenomenon called **outgassing**. Outgassing is the evolution of gases and vapors from the walls of a vacuum system when high vacuum is achieved. In essence, there are always some air molecules remaining in every vacuum system here on earth. The remaining molecules of air, even though few in number, are still colliding with one another and with the walls of the container. These collisions create a small amount of pressure in the system. Even in the highest tech, highest quality research lab in the world, perfect vacuum cannot be attained.

Here are some interesting facts that provide an illustration of the sheer numbers of air molecules that are present in both the ambient air around us and under vacuum conditions. At atmospheric pressure there are approximately 1×10^{19} air molecules per cubic centimeter of volume. A cubic centimeter is a cube that is roughly $\frac{1}{2}''$ x $\frac{1}{2}''$ x $\frac{1}{2}''$. So, per cubic centimeter there are 10 to the 19th power (that's a one with 19 zeros after it - a very big number!) air molecules colliding with one another. If air is compressed to a standard industrial pressure of 100 PSIG (Pounds Per Square Inch Gauge) in a closed system, there are approximately 10 to the 20th power air molecules per cubic centimeter.

At the other end of the spectrum is the rough vacuum world where a very good level of vacuum is 29.9" HgV (Inches of Mercury Vacuum: note that these terms will be explained later) and where perfect vacuum is 29.92" HgV. How many air molecules per cubic centimeter are present at this level of vacuum? 1×10^{-2}? 1×10^{-5}? The actual number is 1×10^{16}! There are still huge numbers of air molecules left in a system at this seemingly high level of vacuum. What is being described here is the nature of vacuum. It does not matter what level of vacuum is reached. In any given system there are always air molecules present and there is, therefore, always some level of pressure in that system. It may be a level of pressure that is far lower than the surrounding atmospheric pressure, but it is pressure nonetheless. Therefore, the definition of vacuum is this:

Vacuum is any pressure in a container or closed system that is less than the ambient atmospheric pressure.

Since there is always pressure in a vacuum system there must be some way to measure that pressure. By understanding how vacuum is measured, the necessary conceptual tools can be developed which will enable us to effectively communicate information about vacuum processes. To get to the root of pressure measurement, it is necessary to go back to the mid 1600's in Italy. A man by the name of Evangelista Torricelli said to himself "I know atmospheric pressure exists and I have to find a way to measure it". And that he did. What Torricelli did was this: He took a hollow glass tube that was open on one end and closed on the other. He filled it with mercury (chemical symbol Hg), tipped it upside down and placed the open end into a reservoir of mercury. What resulted was that the level of mercury in the tube dropped a little bit but remained suspended at a certain height.

There were two primary reasons why the mercury remained suspended in the tube. The first is that atmospheric pressure was pushing on the surface of the mercury reservoir and therefore acted as a force to keep the column suspended. The other reason is a bit more subtle. When Torricelli inverted the tube, the weight of the mercury pulled open a space in the top of the tube. This created a vacuum at the end of the tube. One of the ways to create vacuum is to enlarge a sealed volume. It was the combination of the pull of vacuum in the closed top of the tube and the force of atmospheric pressure pushing on the reservoir that kept the column suspended.

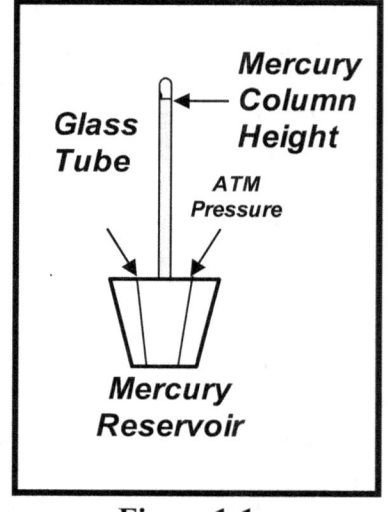

Figure 1-1:
Torricelli's Column

If Torricelli was standing at sea level, the height of the mercury column would be 29.92″. If Torricelli were to walk up the side of a mountain with this apparatus, where the ambient atmospheric pressure was less, what would happen to the column of mercury? The level would drop. The reason it would drop is that there is less ambient atmospheric pressure pushing down on the reservoir. There is, therefore, not as much force available to keep the column suspended. As Torricelli walked up the side of the mountain, the column of mercury would drop lower and lower. If Torricelli could take this apparatus into outer space, which is very close to perfect vacuum, the height of the mercury column would drop to the point where it would be just about even with the level of mercury in the reservoir. When Torricelli developed a way to measure ambient atmospheric pressure, he simultaneously developed a way to measure vacuum.

This now brings us to the point where the five most common scales used to measure vacuum in industrial applications can be introduced. Keep in mind that these are the most common scales used in rough vacuum applications only (0″ HgV to 29.88″ HgV). There are other vacuum applications, specifically medium and high vacuum applications that commonly use other pressure scales. For now though, the focus will be on rough vacuum applications only. Even though some of the conversions presented in this section will at first appear to be laborious, it is critical to understand the material presented here. Not only do all future sections build upon this one, it is also very important to know how to convert from someone else's favorite vacuum scale back to your own personal favorite scale.

If you know and understand the pressure conversions covered here as well as the flow conversions covered in the next chapter, you will have an understanding of the two most fundamental aspects of vacuum science. After learning the material presented in these two sections, you will have the ability to embark on other more interesting and complex vacuum problems.

Table 1-1 has five columns representing each of the five most common vacuum pressure scales. The first row of the table represents atmospheric pressure and the bottom row of the table represents perfect vacuum.

PSIA	Torr	"HgV	"HgAb	Mbar	% Vacuum
14.7	760	0.0	29.9	1,013	0%
13.7	709	2.0	27.9	945	7%
12.2	633	5.0	24.9	844	17%
9.8	506	10.0	19.9	675	33%
7.3	379	15.0	14.9	505	50%
5.9	303	18.0	11.9	404	60%
4.9	252	20.0	9.9	336	67%
3.9	201	22.0	7.9	268	74%
2.4	125	25.0	4.9	167	84%
0.95	49	28.0	1.9	65	94%
0.44	23	29.0	0.91	31	97%
0.21	11	29.5	0.43	15	99%
0.0	0.0	29.92	0.0	0.0	100%

Table 1-1: Common Vacuum Scales

Even though perfect vacuum is not attainable here on earth, we do use it as a reference point when discussing the vacuum scales. The first column illustrates the vacuum (and pressure) scale we referred to earlier which is PSIA or pounds per square inch absolute. At atmospheric pressure the reading on a gauge that measures vacuum in PSIA is 14.7 PSIA. It is called an absolute scale because it uses zero as a base point. In other words, if perfect vacuum could be attained in a system and the vacuum level was being measured with a gauge that reads in PSIA, the reading at perfect vacuum would be zero. Any pressure scale that uses absolute zero pressure as a base reference point is called an **absolute scale**. The PSIA vacuum scale reads 14.7 at atmospheric pressure and as deeper vacuum is attained in a system, the reading becomes lower. Eventually, at perfect vacuum, the reading is 0. In essence, the PSIA scale is a linear scale that follows the pressure inside a system. At half an atmosphere, the reading is 7.35 PSIA which is one half of the atmospheric pressure reading of 14.7. At one tenth of an atmosphere, the reading on the gauge is 1.47 PSIA and so forth.

In contrast to the absolute scales, there are also **gauge relative** scales. This concept is best illustrated by using the familiar automobile tire pressure gauge. When a tire is flat, the reading on the gauge is 0 and as air is pumped into the tire, the reading on the gauge goes up. This scale is called the PSIG scale or **P**ounds per

Square Inch 'Gauge'. The difference between the PSIA scale and the PSIG scale is that the PSIG scale reads zero at atmospheric pressure and goes up from there as opposed to the PSIA scale which reads zero at perfect vacuum and goes up from that point. Since ambient atmospheric pressure can and does change all the time, a PSIG gauge can vary in accuracy whereas an absolute pressure gauge will reflect any changes in atmospheric pressure.

Many compressed air applications use **PSIG** as the scale of choice. When working with vacuum applications, be sure to understand the difference between PSIA and PSIG. The PSIA scale is used in many vacuum applications including hospital, food process and plastics applications. Keep in mind that you can and will find any vacuum scale used in any industry so be ready and able to convert back to your own most comfortable scale when you find others.

The second column of the table is the scale called **inches of mercury vacuum** or **"HgV** as it is symbolized for short. This is THE most common vacuum scale used in rough vacuum applications in the United States. "HgV is not an absolute scale because it starts with and uses atmospheric pressure (top of the scale) as a reference point. In fact, "HgV is the only common vacuum scale that is gauge-relative. "HgV ranges from 0" at atmospheric pressure to 29.92" at perfect vacuum. It is also a linear scale so that at half an atmosphere the reading is 14.96", at one tenth of an atmosphere the reading is about 27" HgV and so forth. The most important thing to keep in mind about the "HgV scale is that even though it is the most common, it is not an absolute scale and cannot be used for flow conversions (which are covered in the next chapter) or any other calculations used in vacuum work.

That covers two of the five columns of Table 1-1: column one, which is PSIA and column two, which is "HgV. The third scale is **inches of mercury absolute** or "HgA as it is also labeled. At atmospheric pressure the reading on this scale is 29.92". "HgA is an absolute scale so at perfect vacuum the reading is 0". It is apparent that there could be some confusion between the "HgV and "HgA scales. They are both exactly the same height and they both begin with letters Hg. When a coworker or customer indicates that a vacuum application is operating at 28" Hg, what should be your first question? Exactly! Is it "HgV or "HgA? Invariably, the answer is a resounding "I don't know!" It is important to know the difference between the two because of the potential for errors in pump sizing or accessory selection. A vacuum application operating at 28" HgA may only require a 2 HP (horsepower) rotary vane or reciprocating vacuum pump, while an application at 28" HgV may require a 75 HP liquid ring or rotary screw vacuum pump.

The good news about the confusion between these two scales is that there is an easy way to find out which scale is being used. When confronted with this situation, ask the question: "what is perfect vacuum on your system or gauge?" If the answer is 30" Hg (most people round the 29.92 to 30), then you know that the scale being referred to is "HgV. If the answer is 0" Hg then you know the scale being referred to is "HgA. It is also acceptable to ask what the gauge reading is at atmospheric pressure or when the vacuum pump is off. If the reading at atmospheric pressure is 30" Hg, then the scale being used is "HgA. If the reading at atmospheric pressure is 0" Hg, then the scale is "HgV. Again, it is very important to know which scale is being used in an application!

Since both the "HgV and "HgA scales are the same overall height and have the same end points (although these points are on opposite ends of the spectrum from one another), it is prudent to be able to convert back and forth from one scale to the other. To convert from "HgV to "HgA or back, there is a very simple formula to use. The conversion formula is 29.92 − "Hg. In other words, subtract the reading on the scale you *have* from 29.92 and the answer will be in the scale that you *want*.

For example, if the specification is 28" HgV and you want to know how many "HgA that is, simply subtract 28" HgV from 29.92 to get an answer of 1.92" HgA. What this is really illustrating is that 28" HgV is the same level of pressure as 1.92" HgA. When some thought is put into this statement, it makes sense. Consulting Table 1-1, it is apparent that 28" HgV is close to the bottom of the table (near perfect vacuum). As perfect vacuum is approached on the "HgV scale, the numbers get higher. The "HgA scale is the opposite. As perfect vacuum is approached on this scale, the numbers get smaller.

As far as vacuum levels are concerned, it follows that 28" HgV is exactly the same pressure as 1.92" HgA. Another example will provide further illustration. How many inches of mercury absolute is 22" HgV? When 22" HgV is subtracted from 29.92, the answer is 7.92" HgA. Again, what this reveals is that 22" HgV is exactly the same pressure as 7.92" HgA. As a final example, the procedure can be reversed so that a conversion can be done from the "HgA scale to the "HgV scale. In this case, the same conversion factor is used in the same manner. How many "HgV is 12" HgA? When 12" HgA is subtracted from 29.92, the answer is 17.92" HgV.

These examples should clearly demonstrate that it is very easy to convert back and forth from one scale to the other. Knowing which scale someone is referring to is the critical part of the problem. As a side note, if you are asked about an application at 15" Hg, it is really not necessary to determine which scale it is because both scales measure about the same level of vacuum at 15" Hg. 15" HgV and 15" HgA are both equal to about one half an atmosphere.

PRINCIPLES IN ACTION

A facilities manager presents you with a vacuum system specification rated at 24.3" HgA and wants to know what the specification would read in "HgV. What is the answer?

SOLUTION:
To convert inches of mercury absolute to inches of mercury vacuum, subtract inches of mercury absolute from 29.92:

29.92 − 24.3 = 5.62" HgV

The previous examples illustrate how to do conversions from an absolute scale ("HgA) to a gauge relative scale ("HgV) and back. It is also necessary to learn how to do a conversion from one absolute scale to another absolute scale. The next example covers this concept. Note again that of the five most popular scales used in rough vacuum applications, all of them are absolute scales with the exception of the most popular one which is "HgV.

For this example, the already discussed PSIA can be converted to "HgA. To convert from PSIA to "HgA, use the conversion factor 2.04. When the vacuum reading in PSIA is multiplied by 2.04, the answer will be the number of "HgA. For example, if a vacuum system is operating at 4.2 PSIA and you want to know how many "HgA that is, simply multiply 4.2 PSIA times 2.04. When the multiplication is completed, the answer is 8.57 "HgA. If you want to go one step further and convert 8.57" HgA to "HgV, simply subtract 8.57" HgA from 29.92 to get a final answer of 21.35" HgV. In this manner, the conversion can be done from PSIA all the way back to "HgV. Also, if desired, the sequence can be reversed to do the conversion from "HgV to PSIA.

For example, if a specification lists a vacuum level of 29.2" HgV and you want to know how many PSIA that is, go through the following steps. First convert 29.2" HgV to "HgA by subtracting 29.2" HgV from 29.92. When this is completed, the answer is 0.72" HgA. Then, reverse operations and *divide* 0.72" HgA by 2.04 to get a final answer of 0.35 PSIA. If you know what units you are starting with, it is an easy conversion to any other vacuum scale. In vacuum applications, sometimes the most difficult information to come by is which scale is being used.

So far, three of the five most common vacuum scales have been covered. This corresponds to the first three columns of Table 1-1. To review, there is PSIA which ranges from 14.7 at atmospheric pressure to 0 at perfect vacuum. Second, there is "HgV which is a gauge relative scale that ranges from 0" at atmospheric pressure to 29.92" at perfect vacuum. Finally, there is "HgA which ranges from 29.92" at atmospheric pressure to 0" at perfect vacuum.

The fourth column in Table 1-1 is the **torr** scale. It is also an absolute scale and has a reading of 760 at atmospheric pressure and 0 at perfect vacuum. The torr scale is commonly used for rough vacuum applications, but it is also used in medium and high vacuum applications as well. The torr scale is based on and named after Torricelli's experiments with the mercury column and can be understood in our day through a simple conversion. There are 25.4 millimeters per inch when the conversion is done between the metric and English scales. If 760 is divided by 25.4, the answer is 29.92. This should demonstrate that the torr scale and the inches of mercury absolute scale are exactly the same except that the torr scale is measured in millimeters and the "HgA scale is measured in inches. The torr scale is also called the millimeters of mercury scale (mmHg) and the two scales are used interchangeably by industry to describe pressure levels in vacuum applications.

In the U.S., the mmHg scale was converted to inches of Hg (and turned upside down in the case of "HgV). The next few examples illustrate how to do conversions from torr to some of the other vacuum scales. To convert from torr to "HgA, simply divide the reading in torr by 25.4. For example, if a specification referenced a vacuum level of 215 torr and you want to know how many "HgA that is, simply divide 215 torr by 25.4. When the division is completed, the answer is 8.46" HgA. Going one step further, this value in "HgA can be converted to "HgV by subtracting 8.46" HgA from 29.92. When the subtraction is completed, the answer is 21.46" HgV.

As another example, 1 torr can be converted to "HgV. First, convert 1 torr to "HgA by dividing 1 by 25.4. After dividing 1 by 25.4, the answer is 0.04" HgA. Then, subtract 0.04" HgA from 29.92 to get the final answer of 29.88" HgV. One torr, or 29.88" HgV is an important dividing point in the vacuum market. An operating vacuum level of one torr is the dividing point between rough and medium vacuum applications. These divisions will be discussed shortly.

SOLUTION:

To convert torr (mmHg) to inches of mercury absolute, divide torr by 25.4. To convert to inches of mercury vacuum, subtract inches of mercury absolute from 29.92:

268 torr ÷ 25.4 = 10.55" HgA

29.92 – 10.55 = 19.37" HgV

One other conversion that can be done with the torr scale is the conversion from torr to PSIA. The conversion factor for this task is 51.71. If the value in torr is divided by 51.71, the answer will be in PSIA. Conversely, if the original reading is in PSIA, that number can be multiplied by 51.71 to obtain a value in torr. As an example, convert 8.5 PSIA to torr. To obtain the reading in torr, simply multiply 8.5 PSIA by 51.71. When the multiplication operation is complete, the answer is 440 torr.

The last of the most common scales and that which completes the final column of Table 1-1 is the **millibar** scale (mbar). Millibar is an absolute scale and is derived from the European Bar scale. At atmospheric pressure the reading on the millibar scale is 1,013 and at perfect vacuum the reading is 0.

One question that is frequently asked is that if one bar is atmospheric pressure and one millibar is one one-thousandth of a bar, why is the reading at atmospheric pressure in millibar 1013? Should it be 1,000 instead? The answer is that the European standard atmosphere is not 14.7 PSIA like it is here in the U.S. In Europe, the standard atmosphere is the Bar which is 14.5 PSIA. The 0.2 PSIA difference between 14.7 and 14.5 requires the use of the "extra" 13 mbar for the U.S. standard atmosphere. The millibar (mbar) scale is used on many types of European manufactured vacuum equipment. Gauges that display vacuum in mbar are supplied on many imported OEM manufactured products such as printing and food process machinery. The millibar scale is also used in many medium and high vacuum applications as well. To do the conversion from millibar to torr, the conversion factor used is 0.7501. If a specification provides a value of 55 millibar for the vacuum level and you would like to know how many torr that is, simply multiply 55 millibar by 0.7501. This will result in an answer of 41.3 torr.

To do the remaining conversions back to "HgV, first divide 41.3 torr by 25.4 to get 1.62" HgA. When 1.62" HgA is subtracted from 29.92, the answer is 28.3" HgV. Hopefully by now, you can see that 28.3" HgV is the same pressure as 1.62" HgA; which is the same pressure as 41.3 torr; which is the same pressure as 55 mbar. In this example, the only conversion left out is PSIA. PSIA can be converted from "HgA like before or it can be converted directly from mbar by dividing 55 mbar by the conversion factor of 68.91. When the division is completed, the final answer is 0.8 PSIA.

It should now be apparent that these five scales are all exactly the same in linear distance. The only real difference is how that distance is divided up. With torr, the full scale is divided 760 times, with millibar it is divided 1,013 times and with inches of mercury it is divided 29.92 times. The exact same physical vacuum gauge can be used to measure any one of these scales. All that is necessary is to take the face plate off the gauge and interchange it with a new face plate that has a different scale printed on it. The new face plate and old gauge will be measuring vacuum accurately to the new scale. The mechanical workings of the gauge do not change, it is only the scale on the face plate that changes.

There are a couple of other items related to the vacuum scales that should be introduced at this time. Sometimes the term '**micron**' is used to describe an operating pressure in vacuum. Microns identify a portion of the torr scale when used as a description of pressure.

Microns are also used as a particle size designation in filtration. Therefore, some confusion can occur over the use of this term. In vacuum work, a micron refers to a micron of mercury. In other words, one-thousandth of a millimeter of mercury. A micron is actually a metric term meaning one-millionth of a meter so it follows that one one-thousandth of a meter is a millimeter (torr) and one one-thousandth of a millimeter is a micron. That is why microns can be used to describe various levels of vacuum. For example, a vacuum application may operate at a pressure of 50 or 75 microns. This refers to 0.05 torr or 0.075 torr respectively.

In essence, since the torr scale is measuring the linear height of a column of mercury in millimeters, it fits that those millimeters can be divided up as finely as we choose. On the other hand, the term micron, when used in conjunction with filtration, refers to an average particle diameter. It still means one millionth of a meter, but in this case it is denoting the average diameter of a particle to be filtered instead of describing a level of pressure. Please refer to Table 1-2 for a summary of the pressure conversions that have been discussed as well as a few others that are commonly used in vacuum applications.

From:	To:	Operation:
"HgA	"HgV	29.92 - "HgA
"HgV	"HgA	29.92 - "HgV
"HgA	Torr	"HgA x 25.4
Torr	"HgA	Torr/25.4
Torr	Mbar	Torr x 1.333
Mbar	Torr	Mbar x 0.7501
Torr	"H2O	Torr/1.868
"H2O	Torr	"H2O x 1.868
Torr	Pascals	Torr/0.0075
Pascals	Torr	Pascals x 0.0075
PSIA	Torr	PSIA x 51.71
Torr	PSIA	Torr/51.71
PSIA	Mbar	PSIA x 68.91
Mbar	PSIA	Mbar/68.91

Table 1-2: Vacuum Conversion Formulas

The terms rough, medium, high and ultra-high vacuum are used when referring to various levels of vacuum. These terms actually designate market sectors within the vacuum industry. It is worthwhile to understand the differences between them and to know when the boundaries are crossed from one to another.

Rough vacuum refers to vacuum systems that operate within the pressure range of 760 torr to 1 torr, or put another way, from 0″ HgV to 29.88″ HgV. Refer to the conversion we did earlier from 1 torr to 29.88″ HgV. These are applications such as hospitals, printing, food processing, material handling, woodworking and plastic thermoforming. The medium vacuum range covers applications from 1 torr to 0.001 torr. Remember that 0.001 torr is one one-thousandth of a torr and can also be called one micron (or one micron of mercury).

Medium vacuum includes applications such as heat treating, lighting, packaging and some automotive processes. High vacuum applications operate in the vacuum range from one micron down to 1×10^{-6} torr (0.000001 torr). These applications include TV picture tube manufacture, specialty metals processing, vacuum deposition and semiconductor applications. Ultra High vacuum applications operate below 1×10^{-6} torr and include space simulation, surface analysis studies and intensive R&D. Many applications, such as heat treating, lighting and semiconductor routinely operate in two or more vacuum regimes.

Rough Vacuum ....760 torr to 1 torr

Medium Vacuum ....1 torr to 0.001 torr

High Vacuum ....0.001 torr to 1×10^{-6} torr

Ultra High Vacuum ....Below 1×10^{-6} torr

It is important to make these distinctions because specific types of vacuum pumps are required to attain vacuum levels within each pressure regime and manufacturers are not always consistent with terminology. For example, a manufacturer may claim that they have a high vacuum pump that can attain 29″ HgV. In this case, they are referring to a vacuum pump that can go to a higher level of vacuum then their other offerings – not true high vacuum. In this case, it is not a reference to a vacuum pump that can attain less than one micron of mercury.

This also brings up the need for an explanation of the terms high and low vacuum. High vacuum refers to very low pressures and low vacuum refers to higher pressures (near atmospheric). For example, a very low or rough vacuum system could be a system that operates at a vacuum level just below atmospheric pressure. Conversely, the phrase 'higher vacuum' refers to a system or process that is operating at a lower absolute pressure than some other comparable system or process. There can be quite a lot of confusion regarding the terms high vacuum, low vacuum, high pressure and low pressure when referring to vacuum applications. The best way to avoid this confusion is to talk in terms of absolute pressures. See Figure 1-2.

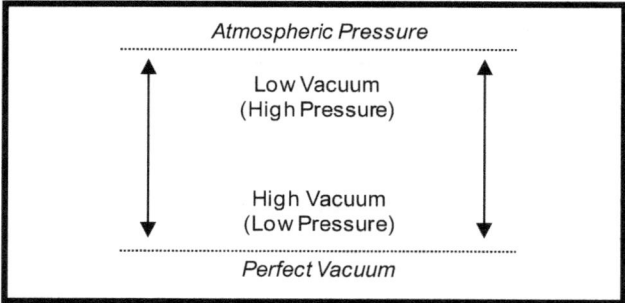

Figure 1-2: High/Low Vacuum

There are more than 30 scales that can be used to describe vacuum in industrial and scientific applications in the U.S. Any one of these scales is just as valid as any other and the trick is to find a scale with which you are comfortable. It is also important to have the ability to convert from any other scale to that one. One trick that can be used if the conversion factor from one absolute scale to another is not known is to take the atmospheric pressure readings of those two scales and then divide one by the other. The number that results is the conversion factor and will enable the conversion back and forth between those two scales. To illustrate, use the atmospheric pressure readings from the torr scale and the PSIA scale. When 760 is divided by 14.7, the 51.71 conversion factor used earlier is obtained. Remember that this only works with absolute scales.

The appendices at the end of this book have several conversion tables available for reference as well as a list of formulas to use when convenient. It might be prudent now to take a break and work several of the problems in Chapter 11 on pressure conversions. The problems are designed to provide a hands-on approach to vacuum and some practice working with the terminology and figures. Remember, for these problems only four basic calculator functions are required: addition, subtraction, multiplication and division. The answers to the quiz questions are at the end of the chapter. Good luck on the problems! When you are finished and are comfortable with pressure conversions you will be ready to move on to the next chapter which is vacuum flow.

Chapter 2

Vacuum Flow

One of the most confusing aspects of vacuum technology to non-vacuum people is the concept of flow in a vacuum system. It is therefore necessary to explain vacuum flow in its most fundamental terms so that an understanding can be built from the ground up. Knowledge of the concepts and terminology regarding vacuum flow will provide you with an intuitive feel for the operation of a vacuum system. Remember that once the concepts of pressure/pressure conversions and flow/flow conversions are understood, much of the remaining material will be easier to comprehend. All other vacuum concepts have their foundation built upon these two fundamentals so it is vitally important that before going on from the first two chapters, you have a complete grasp of these essentials.

To begin with, flow terminology used for rough vacuum systems should be limited to two terms. These two terms are SCFM and ACFM. They will be discussed individually at first and then they will both be discussed in relation to the operation of a vacuum pump. **SCFM stands for Standard Cubic Feet per Minute** and it is the basis for measuring and describing flow at standard atmospheric pressure and temperature. There are several terms commonly used in industry to describe SCFM such as "Free Air", "Atmospheric Air" and "Standard Air" (along with many others). In this text, only the term SCFM will be used to describe mass flow in vacuum systems.

As a rough definition, SCFM can be said to be the mass flow of one cubic foot of air over the time period of one minute at standard conditions. In other words, at 'normal' atmospheric conditions, a cubic foot of air can be weighed at standard pressure and temperature. That weight of air can then be defined as the standard by which to measure other types of flow. That being the case, standard conditions should be defined. In industrial applications, where many of these standards have been developed, standard conditions can loosely be defined as 14.7 pounds per square inch of pressure, 68° F in temperature and 36% relative humidity (water content in the air). For now, relative humidity will be neglected as it has little bearing on this section.

An ideal balloon analogy is the best way to explain the principles of flow under vacuum. An ideal balloon can expand or contract infinitely, the elasticity of the balloon wall is negated and it can never pop or break. If one cubic foot of air at standard conditions is enclosed in an ideal balloon and that balloon volume is pushed (or pulled) through a length of pipe over the course of one minute, the result would be one SCFM of air flow.

It is important to look at the balloon a little more closely though if the nature of air flow is to be completely understood. In essence, what is keeping the balloon at one cubic foot of volume? Realistically, if the balloon walls can expand infinitely, the balloon should continue expanding infinitely due to the air pressure inside. Is it the pressure of the air inside the balloon pushing out or is it the atmospheric pressure outside the balloon pushing in that keeps the balloon at a constant volume? Actually, it is the equilibrium between the two of them: the atmospheric pressure on the outside pushing in is balanced by the internal pressure of the air inside the balloon pushing out.

When this equilibrium is achieved at standard conditions, the result is what is termed one standard cubic foot of air. The air that is inside the balloon can be weighed and when this is done, it is discovered that the weight of one standard cubic foot of air at 14.7 PSIA is approximately 0.075 pounds. It is now possible to examine what would happen if this ideal balloon, with one standard cubic foot of air, is placed inside a vacuum system. When the balloon is placed inside a vacuum system, the volume of the balloon expands.

The reason for the expansion is easy to understand if you think in terms of pressure. The expansion is due to the pressure inside the balloon being greater than the external pressure that surrounds it. When the balloon is exposed to vacuum, the expansion of the balloon volume will continue until the internal and external pressures are once again equalized.

Note that there has been no air mass added to the balloon whatsoever during this expansion. The original air mass is the same as it was before the expansion.

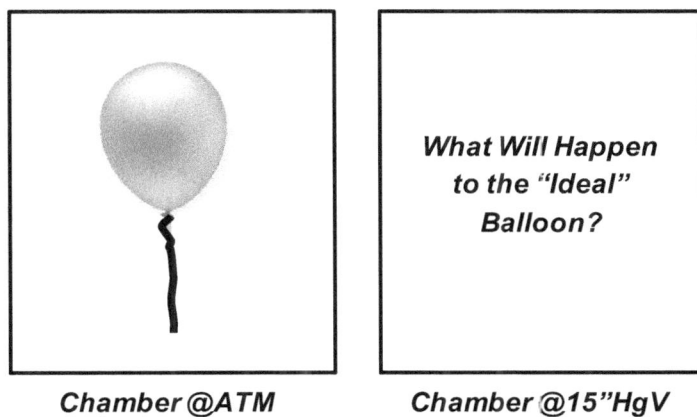

Chamber @ATM   *Chamber @15"HgV*

Figures 2-1A and 2-1B

The new, expanded volume of air in the balloon is the nature of the other type of vacuum flow which is ACFM. **ACFM stands for Actual Cubic Feet per Minute**. It is an increase in the volume of air flow without a corresponding increase in mass of the air flow. Put another way, the density of the air per cubic foot is lower under ACFM flow conditions in vacuum than it is at atmospheric pressure. The 'Actual' part of ACFM stands for the actual conditions within the vacuum system. If the temperature or pressure were changed within the system, the volume of air within the system will change correspondingly. Other industry terms commonly used for ACFM are "Rarefied Air", "Expanded Air", "Aspirated Air" and "Inlet Cubic Feet per Minute" (ICFM). When these other terms are used, they are typically being referred to as ACFM. For our purposes in this text we will use only the term ACFM to describe volume flow.

Now that the general differences between SCFM and ACFM are known, it is possible to examine how these two flow concepts relate to vacuum technology. Back to the 1600's! In the late 1600's, a man by the name of Robert Boyle reasoned that the expansion or contraction of air (and consequently air flow) is inversely related to pressure. As system pressure decreases, the volume of air flow increases. As system pressure increases, the volume of air flow decreases. His theories checked out and became a law of physics called **Boyle's Law**. What Boyle said was that the initial pressure times the initial volume must be equal to the final pressure times the final volume. The initial pressure will be defined as P_1 and the initial volume will be defined as V_1. Subsequently, the final pressure and final volume can be defined as P_2 and V_2 respectively.

Following Boyle's Law, $P_1 \times V_1 = P_2 \times V_2$ or $P_1V_1 = P_2V_2$ as it is more commonly stated. If a system has an initial pressure of one atmosphere and an initial volume of one cubic foot, what would happen to the final volume if the final pressure were reduced by one half? To maintain equilibrium, the final volume would have to double.

Mathematically it looks like this:

1. $P_1 \times V_1 = P_2 \times V_2$
2. 1 atmosphere x 1 cubic foot = 0.5 atmospheres $\times V_2$
3. $1 \times 1 = 0.5 \times V_2$
4. $1 \div 0.5 = V_2$
5. $2 = V_2$, or put another way, the volume doubles when the pressure is cut in half.

If the ideal balloon were placed in a vacuum system at half an atmosphere, the volume of the balloon would double. If the balloon were placed in a vacuum system at one tenth of an atmosphere, the volume of the balloon would increase ten times. If the balloon were placed in a vacuum system at one-hundredth of an atmosphere, the volume of the balloon would increase one hundred fold.

To illustrate:

1. $P_1 \times V_1 = P_2 \times V_2$
2. 1 atmosphere $\times$ 1 cubic foot = 0.1 atm. ($1/10^{th}$ of ATM) $\times V_2$
3. $1 \times 1 = 0.1 \times V_2$
4. $1 \div 0.1 = V_2$
5. $10 = V_2$, or put another way, the volume increases 10 times when the pressure is cut by one tenth.

Next:

1. $P_1 \times V1 = P_2 \times V_2$
2. 1 atmosphere $\times$ 1 cubic foot = 0.01 atm ($1/100^{th}$ of an ATM) $\times$ V_2
3. $1 \times 1 = 0.01 \times V_2$
4. $1 \div 0.01 = V_2$
5. $100 = V_2$, or put another way, the volume increases by 100 times when the pressure is cut by one hundredth.

In every case, the balloon starts with the same one standard cubic foot of air. However, the original one standard cubic foot always expands to greater volumes when the system pressure is lowered. This is the nature of ACFM. *It is the volumetric increase in flow at vacuum conditions without a corresponding increase in mass*. Keep in mind that the only time SCFM and ACFM are equal is at atmospheric pressure. This is because P_1 and P_2 are the same at that pressure. They are different at any other pressure.

It is important to be able to do conversions between SCFM and ACFM in vacuum work. This is because many user specifications are written around flow rates that are stated in SCFM and nearly all vacuum pumps are rated in ACFM. To match a process specification to the correct size vacuum pump, the specification in SCFM must first be converted to ACFM. Only then can the ACFM process specification be matched to the capacity curve for the vacuum pump that is targeted for the application.

The first step in the conversion process is to rewrite the Boyle's Law equation in terms that more closely match industry or process terms. $P_1V_1 = P_2V_2$ can be rewritten in a step by step fashion as follows:

1. $P_1V_1 = P_2V_2$
2. $P_1 \times V_1 = P_2 \times V_2$
3. $P_1 \times SCFM = P_2 \times ACFM$
4. $(P_1 \times SCFM) \div P_2 = ACFM$, or

5. $SCFM \times (P_1 \div P_2) = ACFM$

The original value for SCFM is typically given in the process specification and since SCFM is the rating at standard conditions, it replaces V_1. ACFM is the capacity number we are looking for and since ACFM is the rating at target (actual) conditions, it replaces V_2. P_1, which is always the initial pressure at standard conditions, equates to one of the standard atmospheric pressure readings: 760 torr, 14.7 PSIA, 29.92″ HgV or 1,103 millibar. P_2 is the final or target pressure and it is also given in the process specification.

These particular types of problems involve conversions from flow rates at standard conditions to flow rates at actual conditions, therefore, P_1 is always referenced to atmospheric pressure. Any one of the absolute scales can be used for these conversions, but keep in mind that whatever scale used in the numerator must match the scale used in the denominator. **Also note that ″HgV cannot be used for converting SCFM to ACFM.** The reason for this is because P_1 (atmospheric pressure) expressed in ″HgV is equal to 0. When the number 0 is placed in the conversion equation, the result is: SCFM $\times$ $(0 \div P_2)$ and just does not work. **An absolute scale must be used when doing any SCFM/ACFM conversion.**

Following are a few examples on how to do SCFM to ACFM conversions. These examples are fairly basic so those of you just getting started in vacuum can easily follow along. For more challenging problems, please see Chapter 11 for quiz questions on flow conversions.

EXAMPLE ONE:
Convert the following specification to ACFM: 100 SCFM @ 380 torr.

The best way to keep things straight when doing conversions from SCFM to ACFM is to write out the standard equation with the known variables plugged in and then fill in the blanks. In this example, the standard equation with known variables is:

100 SCFM $\times (P_1 \div P_2) = X$ amount of ACFM.

When the other variables are inserted into their proper places, the calculation can be completed to get the final answer. In essence, the formula comes down to: 100 SCFM × P_1 (which is standard atmospheric pressure of 760 torr) divided by P_2 (which is the target pressure of 380 torr). When the blanks are filled in, the following equation results:

$$100 \times (760 \div 380) = 200 \text{ ACFM.}$$

For this particular application, a vacuum pump with the capacity of 200 ACFM at 380 torr is required.

Just to show that any absolute pressure value will work, PSIA can be substituted for torr. 380 torr is equal to one half of atmospheric pressure and when it is converted to PSIA, the value in PSIA must be equal to one half of atmospheric pressure. The conversion to PSIA from torr is: $380 \div 51.71 = 7.35$ PSIA. The rewritten specification becomes:

Convert 100 SCFM @7.35 PSIA to ACFM.

Working this problem again: 100 SCFM times P_1, which is 14.7 PSIA, divided by P_2, which is 7.35 PSIA. The final equation looks like this:

$$100 \times (14.7 \div 7.35) = 200 \text{ ACFM}$$

As this example clearly demonstrates, it does not matter which absolute scale is used in an SCFM/ACFM conversion as long as units are consistent between the numerator and denominator.

EXAMPLE TWO:
Convert 25 SCFM at 76 torr to ACFM.

Once again, write out the standard equation and fill in the known values. The equation takes this form:

$$25 \text{ SCFM} \times (760 \text{ torr} \div 76 \text{ torr}) = 250 \text{ ACFM}$$

When P_2 (76 torr) is examined, it is found that 76 torr is one tenth of the standard atmospheric pressure of 760 torr. According to Boyle's Law, when the pressure in a system is reduced to one tenth of the original pressure, the volume flow should increase by a factor of 10. In this case it did: $25 \times 10 = 250$. What would happen if the system pressure did not change at all? In other words, can it be proven that at atmospheric pressure SCFM and ACFM are equal? If the preceding example is used and instead of using 76 torr for P2, 760 torr is used for P_2, the resulting equation follows:

25 SCFM × (760 torr ÷ 760 torr) = 25 ACFM

To emphasize the point, the only time SCFM and ACFM are equal is at atmospheric pressure.

PRINCIPLES IN ACTION

A hospital application specification calls out for a vacuum pump that is rated for 159 SCFM at 19″ HgV. How many ACFM is this?

SOLUTION:
First, convert inches of mercury vacuum to an absolute scale. Since ″HgA would be the most direct, subtract from 29.92 to get P2:

29.92 – 19″ HgV = 10.92″ HgA

Then, multiply the amount of SCFM times P_1 and divide by P_2:

159 × (29.92 ÷ 10.92) = 435.6 ACFM

This is a good time to illustrate how SCFM/ACFM flow rates correspond to the capacity of an actual vacuum pump. On the inlet side of a vacuum pump there is ACFM or volume flow. On the discharge side of the vacuum pump, there is what can loosely be described as SCFM or mass flow because the expanded inlet air flow has been compressed back to atmospheric pressure. The SCFM flow on the discharge side of the vacuum pump is said to be loosely described as SCFM because there are a couple of other points to consider.

The first point is that rarely is there *exactly* standard atmospheric pressure on the discharge side of a vacuum pump. Elevation, barometric pressure fluctuations and changes in plant make-up air flow all have an effect on the local ambient pressure. The second point is that vacuum pumps generate some amount of heat during the compression of air back to atmospheric pressure. This results in an increase in discharge air temperature. The discharge air temperatures on some vacuum pumps can reach 300 degrees F or higher. For the discussions on flow in this chapter, discharge air is generally considered to be SCFM and inlet air is generally considered to be ACFM.

When real life applications for vacuum pumps are evaluated, it sometimes takes a very large vacuum pump to pull a small mass flow of air out of a vacuum system. A good example would be a 75 HP vacuum pump that is pulling 15 SCFM out of a vacuum system at a relatively high vacuum. 15 SCFM at atmospheric pressure would only require a fractional HP fan, but at higher vacuum levels, it requires a 75 HP vacuum pump evacuate the same mass of air. Why is this true? The main reason is volume flow. For this example, take an application that has 15 SCFM of air flow going into a vacuum system operating at 12 torr (about 29.5″ HgV for those of you accustomed to using the ″HgV scale). When the SCFM to ACFM conversion is completed, the result is:

15 SCFM × (760 ÷ 12) = 950 ACFM.

What this illustrates about vacuum applications is that pulling 15 SCFM out of a vacuum system at rarefied conditions requires a vacuum pump with a volumetric displacement of 950 ACFM (about 75 HP). In other words, the vacuum pump has to be physically almost 65 times larger than a comparably sized pump at atmospheric pressure to pull the same amount of standard air out of a system.

Conversely, if a 15 ACFM vacuum pump is used on the same application operating at 12 torr, that 15 ACFM vacuum pump would only be pumping out $1/65^{th}$ the amount of air mass at 12 torr as it would be pumping at atmospheric pressure.

That is why large vacuum pumps operating on systems at high vacuum have only a slight amount of air coming out of the discharge pipe. At first glance, it seems like a waste of HP but without a large volumetric capacity vacuum pump, target vacuum would not be attainable.

Temperature

References to temperature have been made several times and it is now a good time to discuss how changing temperatures effect volume flow. Another physicist by the name of Jacques Charles said that changes in temperature will directly affect volume flow. His theory, like Boyle's, was put to the test, survived and subsequently became a law of physics. **Charles' Law states that as temperature increases, gas volume increases and as temperature decreases, gas volume decreases**.

Going back to the ideal balloon example, this can be illustrated very easily. What happens as the ideal balloon heats up? It expands of course. Charles said that the expansion (or contraction) of a gas will occur in proportion to the absolute temperature. A note regarding absolute temperatures: degrees Fahrenheit and degrees Centigrade cannot be used in vacuum calculations. Like the "HgV pressure scale, these temperature scales are relative to temperatures other than absolute zero temperature. That is the bad news – the familiar, customary and conventional temperature scale used by industry cannot be used. The good news is that it is very easy to convert from degrees F to a scale called Rankine by adding 460 to the target value in degrees F. For example, the standard temperature of 68 degrees F is equal to: 68 + 460 = 528 degrees Rankine.

To adjust air flow for changes in temperature from standard conditions, use Charles' Law. Charles' Law formula: $V_1 \div T_1 = V_2 \div T_2$. Like the Boyle's Law formula, this formula can be rewritten with industry standard terms as well. It can be stated that SCFM times T_2, which is the target temperature, divided by T_1, which is standard temperature, equals ACFM adjusted for temperature. After adjusting for pressure with Boyle's Law, it is necessary to make adjustments for temperature changes. The temperature adjustment can be done in one of two ways:

$$SCFM \times (T_2 \div T_1) = ACFM$$

or

$$ACFM \times (T_2 \div T_1) = ACFM \text{ (adjusted for temperature)}$$

As an example, the earlier mentioned specification of 25 SCFM at 76 torr can be used. This specification resulted in a volume flow of 250 ACFM when adjusted for pressure. What would happen to this flow rate if the specification read 25 SCFM @76 torr and $200°$ F? It is not necessary to do the pressure conversion again since the change in pressure is already accounted for. To adjust the ACFM flow for temperature, first find the absolute temperature values. T_1 is the standard industry temperature of $68°$ F which converts to:

$$68 + 460 = 528° \text{ Rankine}$$

and T_2 is $200°$ F which converts to:

$$200 + 460 = 660° \text{ Rankine.}$$

Now that the absolute temperatures have been determined, take the original 250 ACFM, multiply by T_2, which is 660, then divide by T_1, which is 528. The resulting formula and answer are:

$$250 \times (660 \div 528) = 312.5 \text{ ACFM}$$

It is clear that when temperature is taken into consideration, there is an increase in volumetric flow. In this case, the flow increases by 62.5 ACFM. When temperature is given in a process specification, it must be taken into account. Otherwise, there is a potential to undersize the vacuum pump. If the higher temperature was not taken into account for this application, the vacuum pump would be sized for approximately 15 HP. The size of the vacuum pump changes from 15 HP to 20 HP when the elevated temperatures are taken into account. Percentage wise, this is quite a difference in flow!

When both Boyle's Law and Charles' Law are combined, the result is the **General Gas Law**. The General Gas Law states that changes in pressure and temperature effect the volume of a particular quantity of gas.

The General Gas Law:

$$(P_1 \times V_1) \div T_1 = (P_2 \times V_2) \div T_2 \qquad \text{(or } P_1V_1/T_1 = P_2V_2/T_2)$$

Similar to the rewrites on the other two formulas, this formula can be rewritten in terms that are better for our understanding.
The rewritten form is:

$$SCFM \times (P_1 \div P_2) \times (T_2 \div T_1) = ACFM$$

It is now possible to convert SCFM for both pressure and temperature at the same time. To solidify the understanding of this topic, one more example problem should be worked. The new specification is:

Convert 10 SCFM @100 mbar and 120° F to ACFM.

To solve this problem, first write out the base formula with the values for the known variables then fill in the blanks. Millibar is an absolute scale and as you recall, 1,013 millibar is the atmospheric pressure reading on the millibar scale. P_1, therefore, is 1,013. The target pressure of 100 mbar is the value for P_2. To perform the temperature conversion, start with the standard industrial temperature, T_1, which is: $68 + 460 = 528°$ R. The target temperature, T_2, is 120° F which is: $120 + 460 = 580$ degrees R. When these values are combined and placed in the General Gas Law formula, the resulting equation and answer is:

$$10 \text{ SCFM} \times (1{,}013 \div 100) \times (580 \div 528) = 111.3 \text{ ACFM}$$

When the blank formula is written out first and then filled in with process variables one at a time, it becomes very easy to do flow conversions. Always remember to use absolute scales for both pressures and temperatures when doing any flow conversion.

PRINCIPLES IN ACTION

In a food processing application, there is a mixing process that is done with 18 SCFM of air @0.25 PSIA and 210° F. How many ACFM does this equal?

SOLUTION:
First, make sure that absolute pressure and temperature are used. PSIA is an absolute pressure, but 210° F must be converted to Rankine:

210 + 460 = 670 degrees R

The next step is to write out the formula and fill in the blanks:

18 SCFM × (14.7 ÷ 0.25) × (670 ÷ 528) = 1,343 ACFM

There is a derivation on the use of Boyle's Law that can be quite helpful. Suppose there is a hospital application where a 150 ACFM vacuum pump is operating full out and can only attain 18″ HgV in the central vacuum piping system. The maintenance staff has asked for help sizing a vacuum pump that will attain 24″ HgV in the central system. Apparently, some of the doctors are complaining about inadequate vacuum in remote locations. What size vacuum pump should be used?

The answer lies in what is termed the ACFM to ACFM conversion. Unlike converting from SCFM to ACFM like the conversion completed earlier, there is no longer standard conditions at the starting point. Before doing anything first check to see that the current vacuum pump is operating at optimum efficiency and that there are no restrictions to inlet air flow. The inlet isolation valves on some vacuum pumps can be partially closed or there may be cases where there are automatic flow restricting devices (called modulating inlet valves) that are set to close too early. These conditions must be checked and corrected if needed.

Once it has been determined that this vacuum pump is providing full capacity at 18″ HgV, Boyle's Law can be used to calculate the flow that will be required to get them to 24″ HgV. To start with, the pressures that are listed in ″HgV must be converted to absolute pressures. Converting ″HgV to ″HgA would be the most straightforward conversion to use here. Therefore, to determine P_1, take 29.92 − 18″ HgV = 11.92″ HgA. For P_2, take 29.92 − 24″ HgV = 5.92″ HgA. Once P_1 and P_2 are converted to absolute pressures, multiply the ACFM value of the vacuum pump at 18″ HgV by (P_1 ÷ P_2). In this case the formula is:

150 ACFM × (11.92 ÷ 5.92) = 302 ACFM (adjusted).

What this reveals is that to get this particular vacuum system operating at 24″ HgV instead of the current 18″ HgV, a vacuum pump with a total of 302 ACFM would be needed. Note that the capacity of the vacuum pump must be 302 ACFM at 24″ HgV and not just the nominal rating of the pump at atmospheric pressure. This formula can be applied in many cases where the currently installed vacuum pumps are not attaining adequate vacuum.

PRINCIPLES IN ACTION

The maintenance manager for a large printing company is having a problem with his central vacuum system pressure. They have vacuum pumps installed that total 1,850 ACFM and the system can only attain 16.5″ HgV. The production process would run more efficiently at 19.5″ HgV. What is the total capacity required to get them to 19.5″ HgV?

SOLUTION:
First, convert to inches of mercury absolute by subtracting the "HgV values from 29.92:

P_1: 29.92 – 16.5" HgV = 13.42" HgA

P_2: 29.92 – 19.5" HgV = 10.4" HgA

Then, perform the ACFM to new ACFM conversion:

1,850 ACFM × (13.42 ÷ 10.4) = 2,387ACFM

This is a good time to describe the layout of a typical capacity curve for a vacuum pump. A capacity curve can be drawn for any vacuum pump. What this type of curve reveals is that at any given level of vacuum, there is a corresponding amount of delivered ACFM capacity from a particular vacuum pump. The shape of the capacity curve is typically different for each type or technology vacuum pump. A liquid ring vacuum pump will have a different shape capacity curve than a rotary lobed blower and these will have different shaped capacity curves than rotary piston vacuum pumps. It is critical to have access to the capacity curve for a vacuum pump when attempting to place that vacuum pump in an application.

The reason for needing this information is easy to understand. Even though two vacuum pumps may be nominally rated for 100 CFM, the shape and curvature of the individual pump curves can vary dramatically. In the vacuum industry, the term nominal capacity is used to describe the ACFM/SCFM flow at atmospheric pressure. Once vacuum conditions are attained, two vacuum pumps that are nominally rated for 100 CFM can display wildly different flow characteristics. Please refer to Figure 2-2.

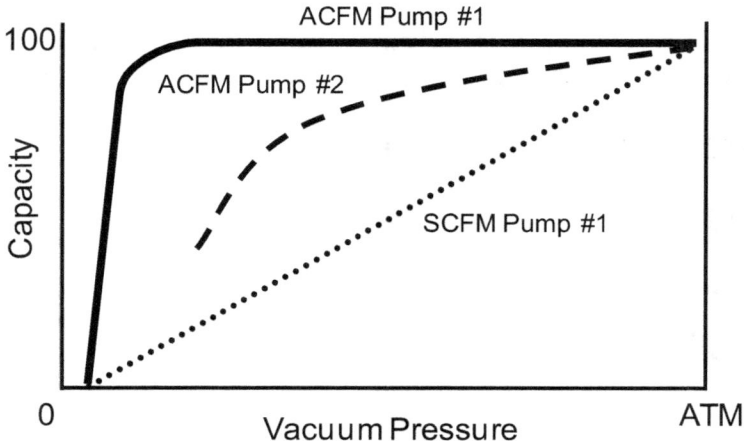

Figure 2-2: Typical Vacuum Pump Capacity Curve

Typically, when reading a capacity curve, pressure will be denoted on the horizontal X axis and flow will be denoted on the Y axis. It is important to read the curve carefully. Most vacuum pump capacity curves are arranged in this way but there can be many variations depending on the industry and the application where the pump is located. Also, some manufacturers develop SCFM curves for their pumps which can look quite different from the corresponding ACFM curves. As a general rule though, most vacuum capacity curves display ACFM as the flow rating of choice.

There are also cases where both the SCFM and ACFM curves are displayed on the same graph. To determine the efficiency of a vacuum pump at a specific level of vacuum, the **BHP** (brake horsepower) curve for that particular vacuum pump must also be available. Many manufacturers will have curves showing both ACFM and BHP on the same page for ease of viewing.

In some instances, it is necessary to find the capacity curve for an already installed vacuum pump. In these cases, it is important to keep in mind that matching the capacity curve with the installed vacuum pump could require some detective work. The pump model and RPM must be consistent with the information listed on the curves. Many pump curves list out an array of pump models with multiple pump **RPM**'s (revolutions per minute) on a single graph. Be sure to identify the correct capacity and BHP curves for the target pump. Also, note that some curves use logarithmic scales while others will use straight linear scales. The same information can be attained from either, but use care when reading the values.

One more note on capacity curves for vacuum pumps. Any one of the five or six most common pressure scales can be used to denote the vacuum level. Sometimes they are used in unconventional ways. For example, there are instances where the mmHg scale (which is the torr scale of course) is used not as an absolute scale, but as a gauge relative scale. Keep a close eye on what values are listed for both atmospheric pressure and perfect vacuum. These provide the surest indication of the scale being utilized and how the curve is arranged.

We have been working with SCFM and ACFM flows in this chapter. There are dozens of other volume and flow conventions used in vacuum applications. For example, there are flow rates designated in liters/minute, liters/second, cubic meters/hour and cubic meters/minute just to name a few. Depending on the industry and application, any one or all of these may be used. The principles for converting from standard conditions to actual conditions do not change. Chapter 11 has an entire section of problems with solutions which will provide practice doing flow conversions. It is important to know how to do both pressure conversions and flow conversion before going on to the next sections of this book.

There is one other mass flow to volume flow conversion that is important to know. This is the conversion from **pounds per hour to ACFM**. In many of the process industries, pounds per hour is used as the standard reference for flow conditions. To properly size a vacuum pump in these applications it is necessary to know how to convert from pounds per hour to ACFM. The formula for the conversion goes as follows:

$$ACFM = (Lbs. \div 60) \times (385 \div MW) \times (760 \div P_2) \times ((460+T) \div 528)$$

Where:
Lbs. = Pounds per hour
MW = Molecular weight of the gas being pumped (air = 29)
P_2 = Target pressure in torr
T = Gas temperature in °F.

The following example provides a quick illustration on how to do these types of conversions.

EXAMPLE:

Convert 78 pounds of air per hour at 120° F and 92 torr to ACFM.

In this case it is a matter of simply filling in the equation with the given variables:

ACFM = (78 ÷ 60) × (385 ÷ 29) × (760 ÷ 92) × ((460+120) ÷ 528)

ACFM = 156.6

These applications are usually a bit more complicated, but it was worth mentioning here how to do a sample conversion. Note that much of the formula consists of components that are already known (i.e. temperature and pressure conversions).

There are conversion tables available in Appendix C for reference as well as a list of flow conversion formulas in Appendix D. It is recommended that you now solve the corresponding problems for this section in Chapter 11. The problems are designed to give you a hands-on approach to vacuum and practice working with the terminology and figures. Again, remember that for these problems you will only need four basic calculator functions: addition, subtraction, multiplication and division. The answers are in the back of the book. Good luck on the problems and when you are finished and very comfortable with pressure and flow conversions, you will be ready to move on to the next chapter which is the pumpdown of closed systems.

Chapter 3

Pumpdown of Closed Systems

Up to this point, many of the concepts and problems have been designed to describe pressures and flows in open vacuum systems. In open systems there is a continuous flow of air coming into the vacuum system from the ambient environment in the form of SCFM or the "free air" which surrounds a vacuum application. There are other cases that require a vacuum pump to be sized for the pumpdown of a closed system. These are applications where there is no additional air being drawn into the system. The air enclosed in a system must be evacuated so that a target level of vacuum is attained in the system within a specified amount of time.

Applications such as pressure treating of lumber, vacuum forming of plastics, vacuum heat treating of metallic parts and vacuum filling of containers are all good examples of this type of process. An application involving the repeated pumpdown of a closed system from atmospheric pressure (or other initial pressure) to a specified vacuum over the course of time is called a **cyclic application**. In contrast, the type of vacuum application where there is a steady flow of air coming in from the surrounding ambient environment is called a **continuous vacuum application**.

Cyclic applications get their name from the fact that the starting pressure is usually the same every time. The goal of a cyclic application is to evacuate a chamber from pressure point A to pressure point B, do whatever work is required, vent back to pressure point A, then repeat the cycle. A faster evacuation of the system to the target pressure is better because the end user achieves greater efficiencies by processing more products per unit time. Typically, the goal in a cyclic vacuum application is to evacuate the system in the shortest amount of time possible.

Even incremental reductions in pump down time can have a significant payback on investment.

Governing the cycle time are the laws of physics and the amount of capital and/or operating expenditures the end user is willing to pay. For example, to evacuate a closed system with a volume of 300 cubic feet to 25 torr in 1 minute would require a vacuum system sized for approximately 75 HP. To evacuate the same system to the same pressure in 10 seconds would require about two 200 HP vacuum pumps. Conversely, to evacuate the same system to the same pressure in 5 minutes would require only a 10 HP vacuum pumping system.

The real question then, is how much will an end-user pay for either the capital expense of purchasing a pumping system or the operating expense of running such a system. The laws of physics come into play when working with pumpdown times that are very fast - less than ten seconds for example. In these cases, the air sometimes just does not want to cooperate and get out of the system as fast as we would want or would expect.

The bottom line in these types of applications is that it usually turns to be one of two questions that must be answered. One, what size vacuum pump would be required to evacuate a system in a given amount of time (solve for capacity), and two, how fast can we evacuate a system with a given size vacuum pump (solve for time). Solving for capacity and determining what size vacuum pump would be required is a good place to start. The formula to determine vacuum pump sizing is:

$S = 2.3 \times (V \div T) \times \log(P_1 \div P_2)$ (solve for ACFM capacity)

S = Pumping Speed (capacity)
V = Volume of the System
T = Pumpdown Time
P_1/P_2 = Initial Pressure and Target Pressure

or

$T = 2.3 \times (V \div S) \times \log(P_1 \div P_2)$ (solve for time)

When the variables are isolated from the equation, it turns out that there are four primary segments: S, V, T, $(P_1 \div P_2)$. Each of these variables should be examined individually. In rough vacuum technology terms, **"S" represents pumping speed** and refers to ACFM capacity. This can get confusing when working with equipment driven by electric motors and pump 'speeds' that are measured in revolutions per minute (RPM). RPM is the common, non-vacuum reference to speed. In other words, RPM measures how fast something is turning.

If pumping speed were referenced to RPM, it would denote that a vacuum pump operates at 1,150 or 1,800 or 3,600 RPM and would be based on the speed of the drive motor (direct drive vacuum pump) or on the sheave ratio (belt drive vacuum pump). In vacuum terms, however, pumping speed means only ACFM capacity. It is one of the idiosyncrasies of the vacuum industry to refer to the delivered capacity of a vacuum pump as pumping speed. Its derivation comes from the concept of throughput, and is beyond the scope of this text. For discussions on the subject in this text, any time the term pumping speed is used, it refers to ACFM capacity.

The next variable is **"V" which represents system volume**. The volume of a vacuum system refers to the entire empty space inside the vacuum system. This includes the chamber, interconnecting piping, accessories like filters, traps and valves, as well as additions to the system like load locks or standby chambers that add physical volumetric space. When the internal volumes of all these items are added together, the result is the entire space that must be evacuated before the intended work can be done.

The best volume unit to use for this particular pumpdown formula is cubic feet. The reason for this is that the final answer should result in a required capacity value of X number of **cubic feet** per minute. If the specifications are in gallons, liters or cubic meters, it is a good idea to convert them into cubic feet. At first glance it would seem that the supply piping does not add a significant amount of volume to a system. In many cases it does, however, so it is important to take the entire piping system volume into consideration when evaluating these types of applications. For example, 25 feet of 3″

diameter pipe does not at first seem to be a significant volume. When the volume of that section of pipe is calculated, however, it comes out to be an additional 1.2 cubic feet. On a 10 cubic foot chamber, this adds about 12% to the total volume. Additional volume will increase the pumpdown time, so it is important to do a careful study of the system so that all component volumes are added into the total.

There is one other item to note on volume. There is a distinction between empty volume and process volume. Empty volume is the volume of the system if measured geometrically. In other words if you measured the inside dimensions of the vacuum system and calculated the volume, this would be considered the empty volume. Process volume, on the other hand, refers to the difference between the volume of the process parts and the geometric volume of the system. This distinction can have significant economic ramifications.

In many vacuum systems, once the target parts are placed inside the vacuum system, the open volume outside of the parts can be very small in comparison to the total volume of the system. The process parts can occupy 90% or more of the geometric volume of the vacuum chamber. If this is not taken into consideration with regards to pump sizing, there is the possibility to oversize the vacuum system for the application. This can result in extra capital and/or operating expense. In contrast, production throughput with an already installed vacuum system can be increased by filling empty voids within the process chamber with non-porous materials so that pumpdown and process times are faster.

The next variable is **"T" which represents time** and just like many other forms of time, manufacturing time equals money. The faster parts can cycle through the manufacturing process, the more cost effective the operation. In this case, the reference is to the amount of time it takes the vacuum pump to evacuate an enclosed system. When working with a specification for the pumpdown of a closed system, the formula units should be in minutes as opposed to seconds or hours. Again, the final answer for pumping speed (capacity) should be in ACFM or actual cubic feet **per minute**.

The last two variables in the pumpdown formula are starting pressure and target pressure. Starting pressure can be either atmospheric pressure or some other level of vacuum where the pumpdown process is started. In the lighting industry for example, there are pumpdown cycles for light bulbs that begin with a starting point below atmospheric pressure. It is important to know the exact starting point because the level of P_1 will affect the pumpdown time to the target pressure. For most cyclic applications operating in the rough vacuum realm, the pumpdown cycle begins at atmospheric pressure. P_2 or target pressure refers to the base level of vacuum required by the process.

With regards to target pressure, many end-users will indicate the need for "30 inches of HgV" or "whatever they can get". Even though it may be difficult to find the specific process information, do not settle for vague specifications like this. Hidden somewhere in many of these cyclic applications is a defined specification that identifies the target pressure. If the application is already in place and running, it is a good idea to actually measure the base vacuum level. In some cases, measuring the vacuum level in a system is the *only* way to find out what the target pressure is for the application. Just like the pressure conversions done in earlier chapters, P_1 and P_2 can be rated in any absolute scale: torr, millibar, inches HgA or PSIA. Any of these scales is acceptable just as long as the numerator matches the denominator.

That completes the description of the four variables necessary to work the pumpdown equation. To summarize, there is "V", the volume of the system in cubic feet, next is "T", the time allotted to complete the evacuation in minutes, "P_1" is the starting pressure, and "P_2" is the target pressure. To calculate "S", which is the required pumping speed, it is necessary to insert the known values into the pumpdown formula and calculate. Two items are important to note at this point. First, if the correct units are used in the formula, the answer obtained with the pumpdown formula will be in ACFM or volume flow.

Second, when sizing a vacuum pump for a cyclic application, it is necessary to have the capacity curve available for that particular pump. If the application requires pumping capacity that is outside the capacity range for a specific vacuum pump, then a larger vacuum pump must be used. The only convenient way to be sure a vacuum pump has adequate capacity for an application is to consult the pumping speed curve for that pump. As an illustration on how to do a calculation to determine the pumping capacity required for a cyclic application, follow along with the example below.

EXAMPLE:
An end-user has a requirement to pull vacuum on a plastic thermoforming system. Thermoforming applications utilize vacuum to pull sheets of hot plastic into molds so that the plastic conforms to the shape of the mold before it cools. Forming plastics this way eliminates bubbles or deformities that would otherwise be trapped on the surfaces of the final product. The system volume is 3,000 gallons, the required pumpdown time is 20 seconds and the target pressure is 28.5" HgV. What size vacuum pump is needed to do the job?

To solve this problem, simply convert the process specifications into terms that are useful in the pumpdown formula. First, convert the 3,000 gallon volume into cubic feet. The conversion is: 3,000 × 0.1337, which equals about 401 cubic feet. Next, take the time of 20 seconds and convert that to minutes: 20 ÷ 60 = 0.333 minutes. Finally, convert the target pressure of 28.5" HgV to an absolute scale: 29.92 − 28.5" HgV = 1.42" HgA. Therefore, the specifications are for a 401 cubic foot chamber that has to be pumped down to 1.42" HgA in 0.333 minutes. For this particular example, it is assumed that all piping and accessories are included in the 401 cubic feet volume specification.

When the values are placed into the formula, the following equation results:

$$S = 2.3 \times (401 \div 0.333) \times \log(29.92 \div 1.42)$$

It is worthy to note at this point that more than just the four basic calculator functions are required to solve this equation. Many calculators, specifically scientific and engineering calculators, will have a log or natural log function available. When 29.92 is divided by 1.42, the result is 21.07. When the log value is taken of that number, it comes out to be 1.32. In effect, the equation simplifies to:

$$S = 2.3 \times (401 \div 0.333) \times 1.32$$

$$S = 3,657 \text{ ACFM}$$

The final answer of 3,657 ACFM is quite a large vacuum pump for a cyclic application such as this. Depending on the pumping technology used, this application could require a 250 HP vacuum pumping system. At this point, realizing the capital expense, the end-user may want to reevaluate the 20 second cycle time. The large capacity vacuum pump needed to pull this system down to the target vacuum level in such a short time is going to be very expensive. If the end-user decided on a one minute cycle time instead of the original 20 second cycle time, what size vacuum pump would be required? Replacing the 0.333 minute time requirement with a 1 minute time requirement yields the formula:

$$S = 2.3 \times (401 \div 1) \times 1.32$$

$$S = 1,215 \text{ ACFM}$$

Since none of the other variables have changed, this calculation is very easy. The final answer of 1,215 ACFM equates to about a 75 HP vacuum pump. As is shown, by increasing the allotted time by a factor of three, the capacity requirement decreases proportionately by a factor of three.

A company that pressure treats lumber has a vacuum chamber that is used to process and treat telephone poles. The process involves pulling a vacuum on the chamber, injecting fluids designed to protect the wood and finally venting the system back to atmospheric pressure. They need to know how fast they can pump the system down to 115 millibar with a 125 ACFM vacuum pump. The volume of the chamber and interconnecting piping is 400 gallons.

SOLUTION:

Since cubic feet are the units in the final answer, the volume of the system in gallons must be converted to cubic feet:

400 × 0.1337 = 53.5 cubic feet

Next, layout the formula, insert the variables and obtain an answer:

$$T = 2.3 \times (53.5 \div 125) \times \log(1013 \div 115)$$

T= 0.93 minutes

There are four considerations to keep in mind when working with closed vacuum systems. The first consideration is the fact that **pressure drop** from the point of use to the vacuum pump must be kept to an absolute minimum. Pressure drop can be an enormous problem in pneumatic systems. Excessive pressure drop can be the cause of poor product quality, excessive energy usage, higher maintenance requirements and nuisance process shutdowns. Piping should be designed to have maximum inside diameters and should be as short in length as possible.

The second consideration, which also involves the interconnecting piping, is conductance loss. Conductance is the actual amount of flow that can pass through a section of pipe at the target pressure. Conductance increases with increasing pipe diameter, so the larger the diameter the better. Any accessories that will be placed in line should be sized for maximum flow and low pressure drop.

The third consideration with closed system pumpdown is placement of the vacuum gauge. If excessive pressure drop is present in a system, attaining the target pressure at the vacuum pump means little if the process pressure is inadequate. Vacuum gauges should be installed at both the vacuum pump and as close to the point of use as possible. The isolation valve between the vacuum pump and the process chamber should be kept as close to the process chamber as possible so that the interconnecting piping is already under vacuum when the cycle starts.

Fourth and finally, leaks can create the need for vacuum pumps with larger capacity than is really necessary. In many cases, when they are not correctly identified, leaks are mistaken for vacuum pump or delivery problems.

PRINCIPLES IN ACTION

The process for degassing and drying metal powder requires vacuum to be pulled on a container filled with metal powder prior to and during a heating cycle. The entire system, which includes a process chamber and interconnecting piping, has a volume of 12.3 liters. A target vacuum level of 7 torr must be attained in 45 seconds or less for the procedure to be profitable. What is the minimum capacity requirement in ACFM to achieve the intended results for this system?

SOLUTION:

Again, cubic feet are the units required in the final answer, therefore, the volume of the system in liters must be converted to cubic feet and the time must be converted to minutes:

$12.3 \times 0.03531 = 0.43$ cubic feet

45 seconds ÷ 60 seconds = 0.75 minutes

Next, layout the formula, insert the variables and obtain an answer:

$$S = 2.3 \times (0.43 \div 0.75) \times \log(760 \div 7)$$

S= 2.7 ACFM

The flow terminology used in this chapter has been mainly ACFM. This is the most common term used in the U.S. to describe flow in rough vacuum applications. There are other flow conventions and terminology used with vacuum systems that have been shipped in from other parts of the world. You can now see the importance of knowing how to do both pressure conversions and flow conversions! Without a working knowledge of these two fundamentals, the focus would be on how to do conversions instead of how to solve problems.

The appendices at the end of the book have several conversion tables available for reference as well as a list of formulas to use at your convenience. It is recommended that you now work several of the problems in Chapter 11 for this section. The problems are designed to provide hands-on practice with calculations involving the pumpdown of closed systems as well as experience working with the terminology and figures. There is a change in the basic calculator requirement though. In addition to the four basic calculator functions, it is now necessary to have a calculator that does log functions. Good luck!

Chapter 4

Vacuum Technologies

It is now a good time to take a break from solving vacuum related problems to discuss some of the different types of vacuum pumps available in the rough vacuum market. Even though these are the most common, there are many different vacuum pumps available and more are being introduced each day. Not only are new pumps being developed, older technologies are being modified to meet new application demands. The most often asked question about vacuum pumps is: what is the absolute best vacuum pump technology available today. The answer, which may seem political at first, is all of them! This answer is based on the wide variations in vacuum applications. There is no one vacuum technology (yet) that suits all of them equally well.

Purchasers and users of vacuum systems have many different reasons and varying priorities when selecting a vacuum pumping system. Some of the prime motivators are delivered capacity, attainable vacuum, types of controls, energy efficiency, price, ease of maintenance, sound level, portability, actual physical size of the pump, etc. The list is endless. Other non-pump related factors such as proximity to a service center, on-time delivery, after sale service and support also play into a users' decision to buy one pump over another. When a single vacuum pump technology becomes the best in all these criteria we can wholeheartedly support that pump. Until then, there will be a healthy mix of technologies from which to choose.

In no particular order, the following six vacuum pump technologies will be discussed in this chapter: reciprocating piston, rotary piston, rotary vane, liquid ring, rotary lobe blower and rotary screw. Each has its place in the vacuum realm and can be found in applications where it is the dominant technology. Again, it should be emphasized that other technologies exist outside of these six, but in terms of an installed base in industrial applications, these are the most popular.

The first technology up for discussion is the **reciprocating vacuum pump**. Recips as they are commonly referred to, look and operate in a similar manner to a piston engine. The piston draws air in from the vacuum system through an intake valve assembly on the down stroke, when it is creating an area of increasing volume. Then, on the upstroke, the piston compresses the same volume of air back to atmospheric pressure and releases it through a discharge valve assembly. These pumps can be oil sealed or dry, and are found in a variety of applications including printing, hospitals and gas well head evacuation.

Performance ratings for these vacuum pumps can be either the actual delivered capacity in SCFM or ACFM or in piston displacement with an associated efficiency value. Pumping capacity efficiencies for recips tend to get lower as higher vacuum levels are attained. As an example, a reciprocating vacuum pump with a piston displacement of 220 CFM could have a 96% efficiency at 5″ HgV. This same vacuum pump operating at 26″ HgV may only have an efficiency rating of 65%. There is a good reason why this efficiency reduction occurs.

The mechanical operation of a recip vacuum pump, with the piston compressing on the upstroke, creates one small problem that becomes magnified at higher vacuum levels. There is a space between the top of the piston on the upstroke and the cylinder head creating what is called a **clearance volume**. At first glance, this would seem to have a minor effect on the operation of the vacuum pump and in terms of mechanical operation this is true. The clearance volume must exist due to the expansion of metal that occurs when the pump heats up as well as to maintain the needed manufacturing tolerances associated with machining pump components. The problem that arises when operating this type of system under vacuum is the air that is trapped within the clearance volume on the piston upstroke is at a pressure that is very close to atmospheric pressure.

As the piston starts its downward stroke, there is a volume of air left in the cylinder. This volume of air works to fill the cylinder volume and helps prevent the intake of new air from the vacuum system. From atmospheric pressure down to about 20″ HgV, this is not a significant factor. In applications where 20+″ HgV is required, however, pumping inefficiencies begin to arise. As deeper vacuum is attained, the inefficiencies become more pronounced. Eventually, when deep enough vacuum is attained, the efficiency goes to 0% and there is no net pumping capacity. At some point, this is true of all other pumping technologies as well.

The high efficiency of the **double-acting piston pump** contrasts the lower efficiency of the single-acting piston pump. The double acting recip typically exhibits higher efficiencies than most other vacuum technologies in terms of ACFM delivered per BHP (brake horsepower) consumed. A double acting recip processes air on both sides of the piston. In other words, there are two sets of inlet valves and two sets of discharge valves per cylinder. These pumps are much less common these days because of high purchase and installation costs. Since there can be significant vibration associated with these pumps, the foundations on which they are installed must be designed for heavy duty service and well engineered.

Controls for recips generally tend to be start/stop via the use of vacuum switches. Continuous operation with some designs at specific vacuum conditions can lead to overheating. Overheating can result in pump and/or seal damage so it is important to apply these pumps properly. However, most designs can run continuously throughout the entire vacuum range and give many years of useful service.

Some of the advantages to using recip vacuum pumps are the ability to do field maintenance, dry compression in many cases (with little or no oil in the compression chamber), rugged cast iron construction and reasonable efficiencies at lower vacuum levels. See Figure 4-1 below for a diagrammatic view of the workings of both single acting and double acting piston vacuum pumps.

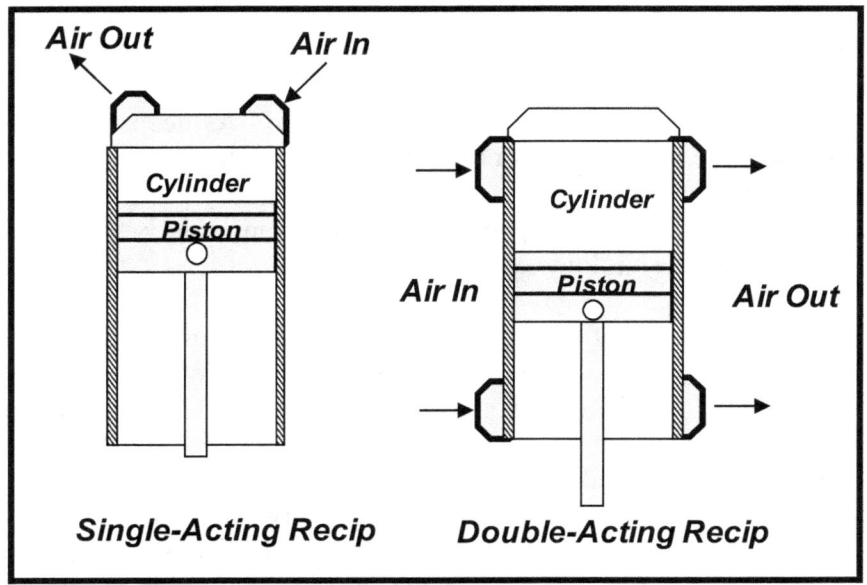

Figure 4-1: Operation of Single and Double Acting Recip.

The next vacuum pump technology is the **rotary piston vacuum pump**. The rotary piston differs from the reciprocating piston pump in that the "piston" is actually mounted in the horizontal position and instead of the piston cycling up and down, this piston moves around sideways in a circular motion. Before reading further, see Figure 4-2 below for a sketch that illustrates the operation of a rotary piston vacuum pump.

You can see from the operating diagram of a rotary piston vacuum pump, that since both the piston and housing are metallic, in order for the pump to work properly the piston and the housing must be very close but not actually touch. There is a film of oil that provides the seal between the piston and housing wall. While the operation of a rotary piston pump is a little bit more complex than the operation of a reciprocating vacuum pump, it is still a very dependable design pump.

There are usually no controls on these types of vacuum pumps other than inlet isolation valves. One of the reasons for this is they are typically placed in applications where full capacity is needed at higher vacuum levels. Since the piston does not come into contact with the housing walls, any particulate that is trapped between the piston and inner housing wall can be pulverized and deposited in the oil reservoir. This makes rotary piston vacuum pumps well suited for applications involving moderate amounts of particulate matter. Just like any closed, oil-sealed system though, too much particulate will eventually wear out internal pump components. The result of this could be the loss of base pressure capability or pumping capacity.

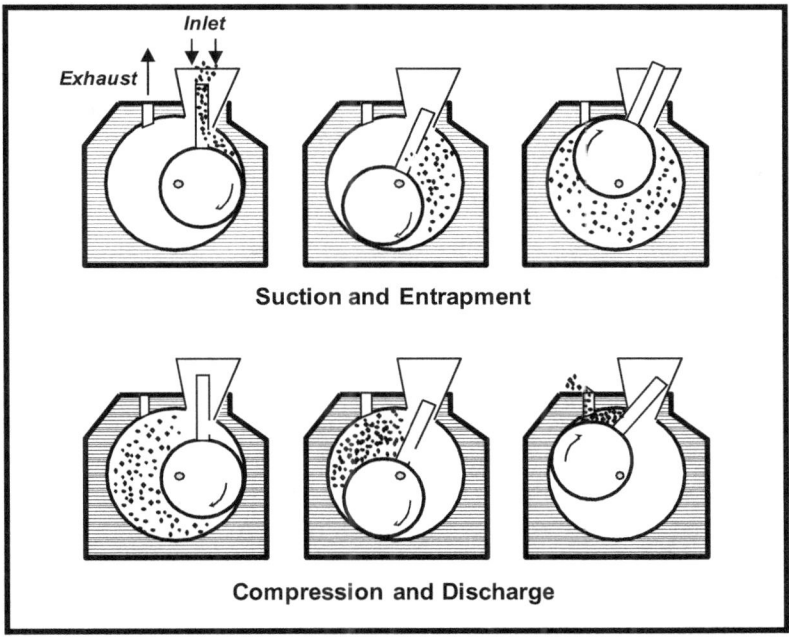

Figure 4-2: Operation of Rotary Piston Vacuum Pump

Rotary piston vacuum pumps can be found in many rugged industrial vacuum applications including heat treating, semiconductor processing, plastics processing, freeze drying and food process applications. One advantage to rotary piston vacuum pumps is very long life. These vacuum pumps are regularly in the 20+ year-old range and in many cases can be serviced in the field if problems occur. Excellent attainable vacuum level is another advantage. Many single stage models can attain a base pressure of 10 microns and two stage models can regularly attain base pressures less than 1 micron (remember that a micron is one one-thousandth of a millimeter of mercury). Rotary piston vacuum pumps are typically larger in physical dimensions than a rotary vane vacuum pump of equal capacity. Due to the momentum of the heavy piston rotating inside the pump, there can also be vibration associated with the use of these types of vacuum pumps.

Next up are **rotary vane vacuum pumps**. Rotary vane vacuum pumps operate by increasing or decreasing a trapped volumetric capacity inside a fixed housing that has an eccentrically mounted rotor with sliding vanes. Also called a sliding vane vacuum pump or just a vane pump, these pumps are small, compact and well suited to many OEM applications where the vacuum pump has to be installed in an enclosure with limited space. In essence, the operation of a vane pump relies on two things: one, that the rotor is mounted off-center (eccentrically) so there are points of larger and smaller volumes in any given revolution; and two, that the vanes slide into and out of the rotor as it turns.

Designs of rotary vane vacuum pumps have vane configurations of two, three, four or multiple. The inlet port is mounted over the area of increasing volume (suction) and the discharge port is mounted over the area of decreasing volume (compression). The rotor is usually directly mounted to and matches the RPM of the drive motor although belt drive designs are common. As the rotor spins, the vanes are pushed outward to the stator wall. Some designs use spring loaded vanes to assist the push against the stator wall while others use centrifugal force.

The following description outlines one rotor revolution in the operation of a rotary vane vacuum pump. As a starting point, picture the first vane in the vertical top center position. The vane is compressed into the rotor but is still making contact with the wall of the stator. As the rotor turns, the vane slides along the stator creating an increasing volume as it moves laterally around the housing circumference and as it slides outward from the center of the rotor. As it passes over the suction port, the increasing volume creates a suction effect and gases are drawn into the compression chamber. Suction continues until the next vane comes into contact with and passes over the suction port.

The air volume is now trapped between the two vanes and the interior of the stator wall. As these two vanes turn, they reach a point of maximum volume where any continued revolution past that point begins to compress the trapped volume. As the trapped volume decreases, the gas is compressed. Compression continues until the lead vane crosses over the discharge port. At that point, the trapped air volume is forced into the discharge port area. When the lagging vane reaches the discharge port, maximum compression is attained and the air is discharged into the exhaust chamber through a discharge valve. The air is then expelled from the vacuum pump.

This cycle repeats itself during each revolution of the rotor and the result is a smooth pull of air from the vacuum system and constant delivery of vacuum to the use point. See Figure 4-3 for a diagrammatic view of the compression cycle in a rotary vane vacuum pump. In oil sealed designs, oil is injected into the compression chamber so that the vane tips slide on a continuous film of oil. By doing this, excellent sealing is attained and vane life is maximized. There are also dry running vane pumps with vanes made of various materials. These materials can withstand the heat and frictional forces generated by the spinning of the rotor.

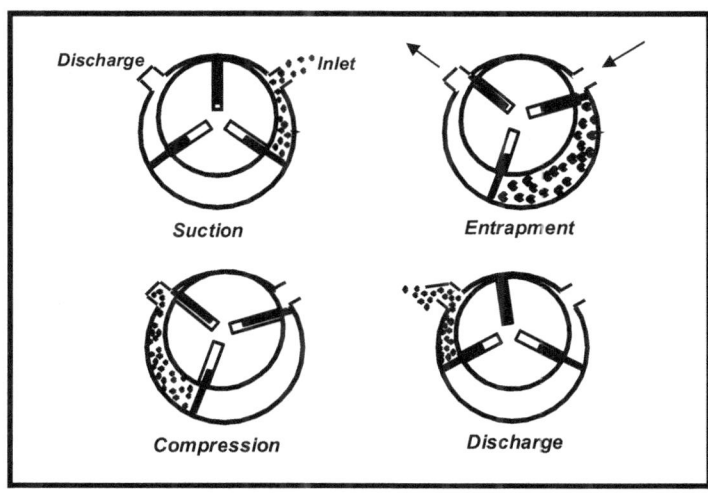

Figure 4-3: Operation of a Rotary Vane Vacuum Pump

Rotary vane vacuum pumps can be found in thousands of different applications. Some of the more common are thermoforming where hot plastic sheets are pulled into the shape of a mold; food processing where mixing or packaging of products takes place; and printing applications where point of use vacuum pumps are mounted on presses, folders and inserters. Advantages to rotary vane vacuum pumps include small overall footprint; smooth, quiet operation; they can be packaged in duplex, triplex and multiplex configurations for a variety of industrial and medical applications; and they typically have good energy efficiencies.

Typical attainable vacuum levels vary depending on the individual design and number of compression stages. As an average, most single stage rotary vane pumps can attain anywhere from 0.05 to 40 torr while two stage configurations can attain vacuum levels of less than one micron. Rotary vane vacuum pumps can be sensitive to particulate matter. When particulate comes between the vane tip and stator, scoring can occur which can reduce the base pressure and pumping efficiency of the vacuum pump. If not cared for properly, rotary vane vacuum pumps can also have shorter life expectancies than some other technologies.

It is important to note that with any closed oil system vacuum pump, it is critical to understand what is being drawn into the pump in the form of particulates, gases, liquids and vapors. Many different types of vacuum pumps will work in an application, but some technologies require more protection than others. If the process gas stream is completely understood, steps can be taken to rectify any potential pump-life shortening problems before they occur. When these steps are taken, pump and process up-times are maximized.

Keep in mind that any vacuum pump can be protected from any process. The question is whether or not it is worth paying for protection when other, more suitable alternatives exist. The bottom line is that in any vacuum application, it is necessary to know exactly what is entering the pump. With this information, informed decisions can be made on which technology, pumping system or protection to install in that particular application.

Now back to the discussion of vacuum technologies. The next pumping technology is the rotary **liquid ring vacuum pump**. In principle, it is very similar in operation to a rotary vane vacuum pump in that the internal compression chambers are sectioned off between each other with individual walls or "blades" that move in a circular motion. Instead of vanes, however, there are impeller blades that move into and out of a liquid seal. The seal medium can be water, oil or any one of a large number of other potential seal fluids. The heart of the liquid ring vacuum pump is the impeller.

Looked at from the end, the impeller appears similar to a paddle wheel on a paddle boat. From the center hub, each "spoke" is equally spaced around the perimeter. The impeller is housed in the impeller housing and is offset from the exact center (eccentrically) so that as it turns, there are places where the seal liquid is very close to the center hub and opposite those places, the seal liquid is at its furthest point away from the center hub. See Figure 4-4 for a view on the operation of a liquid ring vacuum pump.

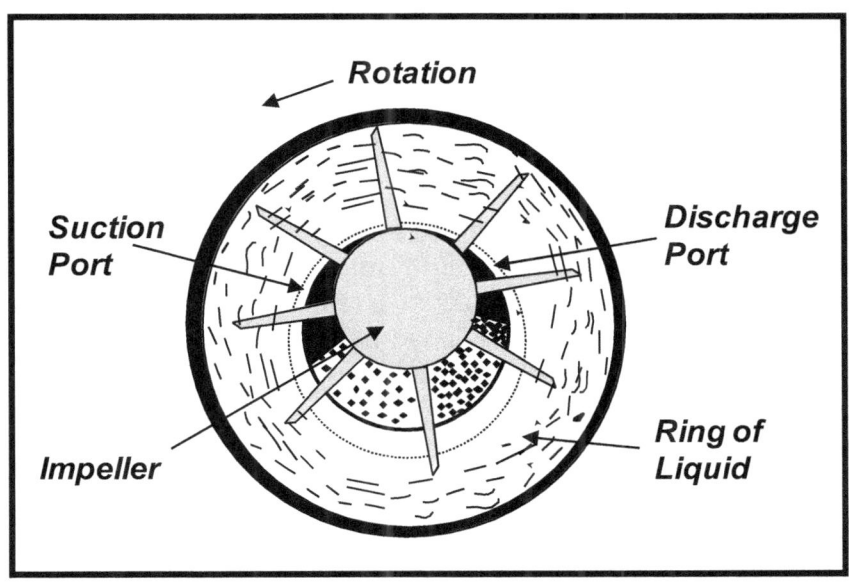

Figure 4-4: Operation of a Liquid Ring Vacuum Pump

The manner in which a liquid ring vacuum pump operates is that as the impeller rotates, the impeller blades pass over a suction port where the increasing volume creates a suction effect. Similar to the rotary vane design, there is entrapment, compression and discharge of the air as the impeller rotates and the individual chambers cross the suction and discharge ports. Sealing fluid is injected into the impeller housing at a prescribed rate so that as the impeller turns, it forces the injected liquid to the outermost portion of the circle formed by the impeller housing around the impeller. This spinning motion creates what is called the liquid ring.

Since the impeller is offset from the center of the housing, there are places where the liquid is close to the center hub, creating a very small volume, and places where the liquid ring is much further away from the center hub, creating an area of larger volume. As the impeller rotates, the impeller blades actually move into and out of the liquid ring so that the sealing liquid becomes a "liquid piston" and forms a wall of the compression chamber. The other two walls of the chamber are the adjacent impeller blades. As the impeller turns over the suction port, the volume of the compression chamber is increasing. This creates a suction effect and draws air in from the pump inlet.

By the time the chamber volume is maximized, it is no longer over the suction port and the volume of air is trapped. As the impeller continues its revolution around the inside of the impeller housing, the liquid piston begins to close in. The decreasing volume compresses the trapped air inside it. Eventually, the compressed air comes in contact with the discharge port and is expelled from the vacuum pump. It is at this point that the liquid ring is just about in contact with the actual hub itself and the chamber volume is near zero. Onward the impeller rotation continues until it again comes into contact with the suction port.

This is only one cycle of the many chambers in a liquid ring vacuum pump. Since there is a continuous supply of chambers, there is continuous suction at the inlet port and continuous air discharge at the exhaust port. In other words, there is a smooth, continuous evacuation of air from the system.

Liquid ring vacuum pumps have a number of design characteristics that make them suitable for tough applications. Liquid ring vacuum pumps do very well in applications where there is a high particulate load. In most cases, the bearings that support the drive shaft and impeller are located in a separately greased chamber. This means that the bearings do not come into contact with any gas or particulate generated by the process. Therefore, pump life is not typically shortened due to bearing failure. Also, there is no contact between the impellers and the impeller housing so there is typically no mechanical wear on the impeller. Liquid ring pumps can be made of a variety of materials that resist process chemical attack. This keeps corrosion of pump components to a minimum.

Liquid ring vacuum pumps are very widely used and can be found anywhere vacuum is required. Typical applications where liquid ring pumps are installed are chemical processing, poultry processing, wood working, soil remediation, hospital central vacuum and plastics manufacturing. Depending on the configuration and the type of seal liquid used, single stage liquid ring vacuum pumps can attain 20″ to 25″ HgV. Two stage liquid ring vacuum pumps can attain vacuums to 29+″ HgV.

Liquid ring vacuum pumps can be packaged as once through, or open, systems or as stand-alone closed loop systems. Open systems typically use water as the sealant in once-through or partial re-circulating configurations. Packaged, closed loop systems that use oil as a sealant have become very common in industrial applications over the last few years. Just like in oil sealed rotary vane vacuum pumps, it is important to have proper inlet filtration prior to the vacuum pump in heavy particulate applications.

Rotary lobe vacuum pumps are next on the list. Rotary lobed pumps, also called "Roots" pumps, have been around for many years and are used in many tough industrial and chemical applications. Rotary lobed vacuum pumps are generally considered to be dry vacuum pumps since they typically use no liquid sealant to seal the clearances between the lobes and the inner housing. In vacuum applications, rotary lobed vacuum pumps are used in two general ways. First and most common, they are used as the primary vacuum pump in a system. Second and less common to rough vacuum applications, they are used as "boosters" to other vacuum pumps. This is the reason rotary lobe vacuum pumps are sometimes referred to as booster pumps or blowers.

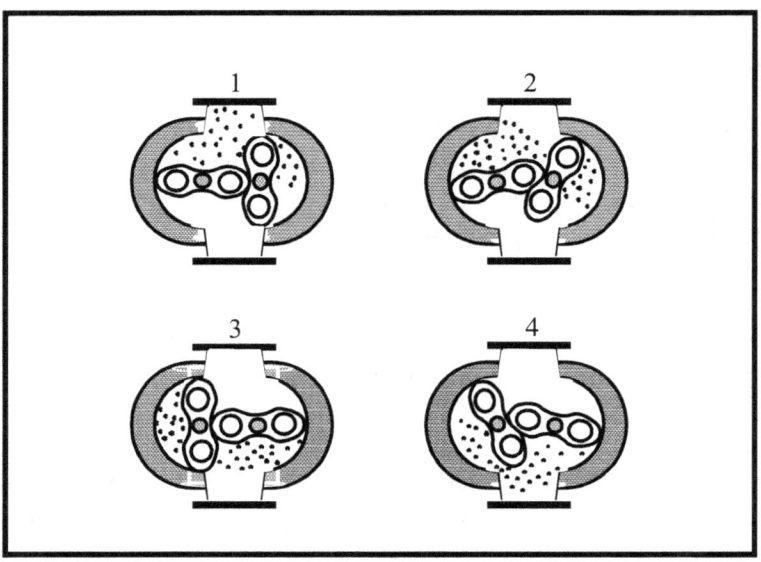

Figure 4-5: Operation of a Rotary Lobe Blower

When viewed from the side, a rotary lobe vacuum pump can be described as two figure eight lobes that turn in opposite directions inside a pump housing. For each revolution of the lobes, there are several pockets of air that are drawn into and through the pump. Clearances between the tips of the lobes and inner pump housing are very close and are manufactured that way to reduce slippage of air back into the vacuum system around the lobe tips. See Figure 4-5 for a diagrammatic view of the operation of a rotary lobe vacuum pump. Rotary lobe vacuum pumps have a very low built-in compression ratio and when used in single stage configurations, can attain only moderate vacuum levels. These pumps can be staged, however, so that greater compression ratios are attained and deeper levels of vacuum can be reached.

Some pumps use water or other seal fluids injected at the inlet of the pump to assist in sealing or cooling. These liquid sealed pumps can attain higher levels of vacuum than their dry counterparts. For the most part, single stage, non-injected rotary lobe type vacuum pumps can attain 15″ HgV to 18″ HgV and are used in many material handling and conveying applications. Dry type vacuum pumps like these are used because contamination of the compression area with particulate usually does not damage the vacuum pump to any great degree. Material handling applications can generate quite a lot of particulate matter so these pumps are a natural fit in those applications. Dry compression also has a big advantage when considering the costs to dispose of contaminated lubricants.

Sometimes two or more rotary lobe vacuum pumps are **staged** in series. The lead pump is connected to the vacuum system and discharges into the inlet of the second pump which discharges into the inlet of the third pump and so on. These systems have greater compression ratios because of the staging arrangement and can attain much higher vacuum than an individual rotary lobe vacuum pump. These configurations are not as common and warrant special considerations to eliminate condensation and the heat generated by compression.

The other typical use for a rotary lobe vacuum pump is as a **booster** to another vacuum pump. In these systems, a liquid ring, rotary piston or rotary vane vacuum pump is used as a "backing" pump to the booster pump. The booster is the lead pump connected to the vacuum system. These systems typically operate at the low end of rough vacuum and well into the medium vacuum range (below 1 torr). The booster pump combination provides a substantial capacity increase over an individual backing pump and

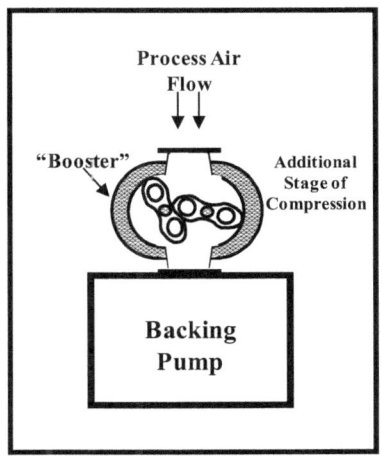

Figure 4-6: Booster Package

raises the efficiency of the pumping system when used properly. In

61

these cases, since the booster provides high capacity, the backing pump can be downsized. This type of pumping system arrangement saves on capital costs and floor space. These systems are used throughout industry in applications such as lighting, vacuum furnace, semiconductor processing and food processing. See Figure 4-6 for a drawing of a booster type vacuum system.

Rotary lobe blowers have many advantages in industrial applications. The fact that they provide dry compression is a great asset when they are used in high particulate or chemical applications. Also, rotary lobe pumps are very efficient when used within their design pressure ranges. When higher vacuum levels are required, however, individual rotary lobed blowers can be limited in the amount of vacuum they can draw. Silencers and other sound attenuation devices are commonly used to help reduce the sound generated by these types of vacuum pumps.

The last vacuum pumping technology to be discussed is the **rotary screw vacuum pump**. Rotary screw vacuum pumps were commercially developed in the early 1980's and are one of the newer technologies on the market. The principle of operation is identical to that of a rotary screw air compressor and, in fact, these pumps were developed from rotary screw air compressor designs. Two helical screws, one male and one female, are intermeshed within a rotor housing. These three components (the two helical screws and housing) along with the associated fittings, bearings and end-caps are commonly referred to as the "**airend**".

At one end of the airend is the inlet port where air is drawn in and at the other end is the discharge port where the compressed air is discharged back to atmospheric pressure. Refer to Figure 4-7. The rotors turn in an outward motion (looking at the rotors from the end). As the rotors turn, pockets open up and air becomes trapped between the rotors and the housing wall. As these pockets move toward the discharge port, compression is taking place as the air is being forced into smaller volumes. Once the rotors open over the discharge port, the pockets close and air is forced out because the male and female rotors are once again in contact. The cycle repeats itself continuously for as long as the rotors are turning. Oil is typically injected into the airend to assist in sealing, lubrication and to help carry away the heat of compression.

Rotary screw vacuum packages are very similar to oil sealed liquid ring vacuum pump packages in overall package layout. They both require oil reservoirs, air/oil separators, coolers and electrical control panels. Rotary screw vacuum pumps have been used in many industrial vacuum applications such as printing, plastics, food processing, wood working and hospitals. Typical ultimate vacuum levels are anywhere from 29.5″ HgV to 29.9″ HgV, however the vast majority of applications for rotary screw vacuum pumps operate in the vacuum range between 15″ HgV and 28″ HgV.

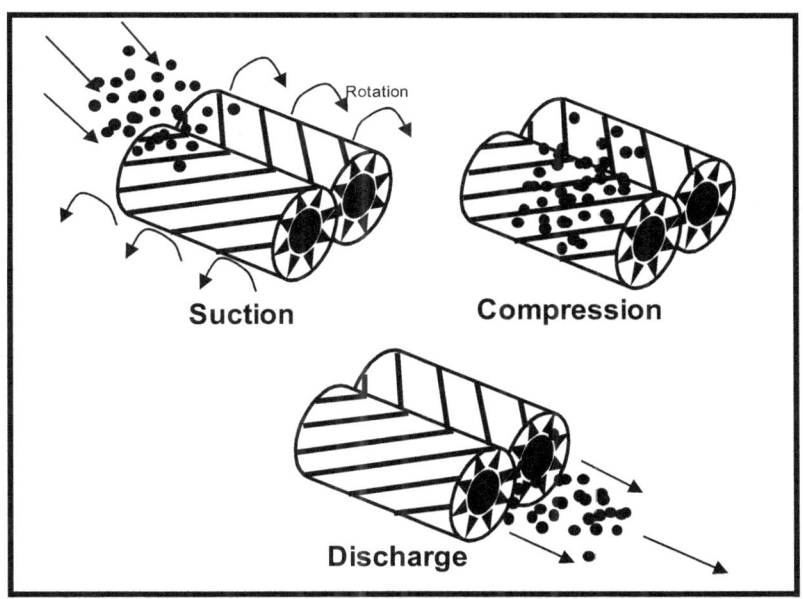

Figure 4-7: Operation of Rotary Screw Airend

One of the advantages of rotary screw vacuum pumps is long life. Typically, bearings used in rotary screw airends are designed for and constructed for air compressor applications. Many of these same airends are used for air compressors in 100 PSIG applications. When pump lubricants are well maintained, rotary screw vacuum pump life is very long. One of the other advantages of rotary screw technology is the **modulating inlet valve** which is a carryover from air compressor applications. The modulating inlet valve is a device that is designed to control system vacuum level automatically. The user sets the range by adjusting the high and low set points and the modulating inlet valve maintains the range automatically by opening or closing in response to system demand for vacuum. The opening or closing of the valve allows for more or less air mass to enter the pump thereby keeping the vacuum pump at its most efficient operating point automatically. As mass flow increases or decreases, horse power increases or decreases in response. The other obvious benefit is reasonably accurate process control built right into the pump. Rotary screw vacuum pumps typically have good CFM/BHP efficiencies.

When choosing a vacuum pumping technology for an industrial application, it is important to be armed with pertinent information regarding the vacuum pump. Items like capacity curves, BHP curves, maintenance manuals, specifications and installation guidelines are all important pieces in the quest to solve the technology puzzle. It is important to know exactly what will be going into the vacuum pump so the proper vacuum pump or necessary protection devices can be selected for that particular application.

Chapter 5

Vacuum Inlet Filtration

One of the most important considerations in any vacuum system is the protection of vacuum pumps from particulate contamination. Given the high cost to purchase vacuum equipment, it is wise to protect that investment with a highly effective and relatively low cost inlet filtration system. There are many examples of installations where expensive vacuum pumps have been installed in production systems only to have them fail prematurely due to inadequate inlet filtration. Of the many reasons to protect a vacuum pump, the primary reason is to lower or prevent capital costs. The initial investment in a vacuum pump can be anywhere from a few hundred dollars to several hundred thousand dollars. The purchase cost for inlet filtration to protect this investment is typically less than 10% of the initial investment made for the vacuum pump.

Even though the overall cost for inlet filtration is low, many companies will opt for less than adequate protection citing additional maintenance costs or added system complexity as reasons to defer. As we will see, these are not reasonable assumptions and can be readily proven incorrect. Beyond protecting capital expenditures, reasons to protect vacuum pumps include lower overall maintenance costs, less production downtime (which is typically the highest but least defined cost) and fewer nuisance problems associated with plant production machinery.

Before delving into the nuts and bolts of inlet filtration, it is prudent to look at the specific financial reasons for not only having inlet filtration, but also keeping it in top condition. Since costs for maintenance time can vary widely, only component cost will be evaluated in this example. Table 5-1 illustrates the costs associated with purchasing and maintaining a 300 CFM vacuum pump over the course of 15 years both with and without an inlet filter. As a very conservative assumption, the vacuum pump should have a normal life before rebuild of 5 years with adequate filtration and a shortened life of 2 years without adequate filtration.

Pump Maintenance Costs		
Year	w/o Filters	w/Filters
1	$12,000	$12,600
2	$6,000	$180
3	$0	$180
4	$6,000	$180
5	$0	$6,180
6	$6,000	$180
7	$0	$180
8	$6,000	$180
9	$0	$180
10	$6,000	$6,180
11	$0	$180
12	$6,000	$180
13	$0	$180
14	$6,000	$180
15	$0	$6,180
	$54,000	$33,120
	Difference:	$20,880
	Investment in Filtration:	$3,300

300 CFM Vacuum Pump ($12,000)
Inlet Filter 5% cost of Pump ($600)
Two Filter Elements Annually @ $90 each
Repair Cost: 50% Cost of New Pump

Table 5-1: Maintenance Costs w/ and w/o Filtration

To illustrate the conservative approach used here, in many applications vacuum pumps are rebuilt on a semiannual basis due to failure of the inlet filter or lack of an adequate inlet filtration system. As the table clearly illustrates, over the course of 15 years with a total investment of $3,300 in filters and replacement elements, there is a minimum return of $20,880. The findings in this table assume that the vacuum pumps can be rebuilt after failure due to contamination. If a catastrophic failure occurred and the vacuum pumps could not be rebuilt, complete replacement would be necessary. This, of course, would add significantly more money to the "return" row. When maintenance time is considered, there is still a big plus in the "with filtration" column because it typically only takes several minutes to change out an inlet filter, while it could take several hours or even days to rebuild or replace an entire vacuum pump. The bottom line is that it is far more advantageous to use proper inlet filtration.

After evaluating the dollar savings that can be realized with a good inlet filtration system, it is time to more closely review the "why" and "how" of inlet filtration. First, it is prudent to examine what specifically happens when particulate enters a vacuum pump and why vacuum pumps fail when excessive particulate is present. Given the wide variety of vacuum pumping technologies, it is not possible to examine every failure mode for each type of pump. Instead, conclusions can be drawn from evaluating several general failure characteristics that are common to some or all vacuum pumps. Keep in mind that any vacuum pump can exhibit any one or all of the listed failure characteristics. Also note that some vacuum pump technologies are more resistant to particulate contamination than others. The amount and type of particulate contamination generated by the process will usually determine the severity of the failure.

Bearings: When vacuum pump bearings come into contact with a contaminated process air stream or with lubricant/seal fluid that has been in contact with a contaminated process air stream, there is the potential for bearing wear. Depending upon their location and the overall design of a pumping module, bearings can be very sensitive to particulate contamination. Some pump manufacturers design their pumps with heavy duty bearings so particulate contamination will take much longer to manifest itself in these pumps. Thrust loading, operating temperature and rotational speed all have an impact on bearing wear and subsequent bearing life. When particulate contamination is present in the fluids that lubricate bearings, wear is accelerated due to increased friction and actual etching of bearing surfaces and bearing races. Bearings operate at their highest efficiency and longest life when they are in perfect alignment. Particulate wear changes the alignment and allows for 'play' which increases surface temperatures and accelerates the degradation process.

Operating Temperature: There are four general ways in which particulate contamination increases operating temperatures in vacuum pumps. The first and most obvious is physical restriction of components within the vacuum pump itself. As internal components become caked with debris, additional energy is required to overcome added frictional resistance.

Second is the physical blocking of oil passages within the vacuum pump. When particulate contamination builds in oil passages, there is a restriction in the amount of oil that can pass through these passages. Oil channels in coolers or strainers may become completely blocked. On some pump designs, oil is used as a medium to carry away the heat of compression. When not enough oil is allowed to pass through the pumping module, pump operating temperature increases. If the operating temperature becomes high enough, lubricant failure occurs which is certain death for a vacuum pump. The third way particulate contamination increases pump operating temperature is by partially blocking the discharge port. The discharge port on a vacuum pump is that point where the air and/or air-oil mixture is ejected from the pumping module. If normal flow through this port is restricted, back pressure will build. Back pressure causes an increase in drive motor horsepower resulting in an increase in pump operating temperature.

The fourth method is a bit more subtle. On dry-compression type vacuum pumps, particulate can actually pass directly through the vacuum pump and end up in the ambient air inside an equipment room. This dust will blanket everything in the room and reduce the heat rejection capacity of drive motors, coolers and the vacuum pump itself. The result is in elevated pump temperatures.

Component Wear: Particulate abrasion can cause many problems in a vacuum pump. Over time, the presence of a continuous stream of particulate matter within the process air stream or within the pump lubricating/seal fluid will wear down internal vacuum pump components. This results in vacuum pump failures, reductions in pumping capacity and loss of attainable base pressures. These effects can be more or less severe depending on the particular type of vacuum pump.

Inlet Accessories: There are instances where excessive particulate contamination will hinder the operation of vacuum pump inlet accessories. Isolation valves, check valves, modulating inlet valves and other components are all subject to improper functioning as the result of excessive particulate contamination. In some cases, the combination of particulate and water contamination will create a condition where a component completely freezes up. These components will have to be disassembled, cleaned or completely replaced.

Remember that there are many different kinds of contamination that can enter a vacuum pump. Everything from the process travels from the point of origin, through the vacuum piping, through any inlet accessories and finally into the vacuum pump. It is important to know exactly what kind of contamination is present so that the proper precautions can be taken to capture it prior to the inlet of the vacuum pump.

As a first step, it is important to understand conceptually that even though a system is under vacuum conditions, there is still the potential for the incoming air stream to pick up and carry particulate matter from the process and/or piping back to the vacuum pump. Given that this text is written primarily for vacuum applications above 1 torr, there is a formidable amount of mass flow in most vacuum systems. Going back to the SCFM/ACFM flow calculations from earlier chapters, we can calculate how much SCFM is flowing through a vacuum pump (and therefore the vacuum piping). As an example, a vacuum pump with a nominal capacity rating of 150 ACFM can be evaluated at varying levels of vacuum. Table 5-2 outlines the SCFM flow rates for a 150 ACFM vacuum pump at several different pressures.

It's not until approximately 28″ HgV that the flow rate is under 10 SCFM. Depending on the pipe diameter and subsequent **air velocity** (hence the ability for the air stream to pick up debris), the air mass may or may not be high enough to pickup or move a significant amount of particulate matter. Also, as a general rule of thumb, note that at 15″ HgV there is roughly half the nominal capacity of the pump in SCFM. The closer the system is to ambient pressure, the higher the capability (and likelihood) to pick up debris. Cyclic applications that start at atmospheric pressure and pull vacuum to low target pressures have the greatest opportunity to pick up contamination and bring it back to the vacuum pump. The reason for this is that cyclic applications start at atmospheric pressure where there is the highest amount of air mass flow.

Nominal Pump Rating: 150 ACFM	
Vacuum ("HgV)	SCFM Flow
0	150
5	125
10	100
15	75
20	50
25	25
28	10
29	5
29.5	2
29.9	0

Table 5-2: Mass Flow Under Vacuum

When considering an inlet filter for a vacuum application, the first couple of questions that must be answered are what type of media should be used and what are the correct pipe connections for a particular model vacuum pump. Unless you have a process engineer in your back pocket that can tell you exactly what size and type particulate will be generated and what process gases are present in that application, the first question can be difficult to answer.

Questions regarding connection size are easier to answer because in most cases, filter connection sizes will match those of the vacuum pump. For a large number of installations, these two questions are not answered adequately. This results in the placement of inlet filters that have marginal or no capacity whatsoever to do the job. In this chapter, many of the fundamental methods for determining proper inlet filter size will be discussed. Practice of these methods will enable you to confidently size and install inlet particulate filters for most standard vacuum applications.

Before getting into the specifics of determining the proper inlet filter, it is important to learn a few basic terms that are used in the filter industry. The first term is the **micron**. A micron is a linear measurement that equates to one millionth of a meter (or one thousandth of a millimeter). In inches, a micron is equal to 0.000039" or put another way, $1/25,400^{th}$ of an inch. Even though the micron highlighted in earlier chapters with regards to pressure and this micron mean basically the same thing, they are applied a little bit differently. In filtration terms, the micron is the standard designation for describing the average diameter of a particle that is separated via some type of filter media. To provide an illustration of the relative sizes of particles in this range, human hair has an average diameter of approximately 80 to 120 microns. The smallest particle visible to the human eye is normally in the 10 to 40 micron range.

Designations such as 1 micron, 5 micron, 10 micron, 50 micron and 100 micron are used to describe the capabilities of filter media. What these size designations mean is that each particular media rating has a stated efficiency for filtering particulate. For example, a 10 micron filter media may have a 99.9% removal efficiency for all particles that are 10 microns and larger. What that means is that a 10 micron filter media can and will filter larger particles to that efficiency level.

What is not stated but is true, is that the 10 micron filter media also has an efficiency rating for smaller particles as well. The efficiency rating will be lower for the smaller particle sizes but still may be good enough for an application. As the particle size gets smaller, efficiency becomes lower and eventually the efficiency is so low that the filter media is unusable. To illustrate how this works, please refer to Figure 5-1 below. Figure 5-1 is a typical particle efficiency curve for a example filter media.

Typical Filter Efficiency Curve

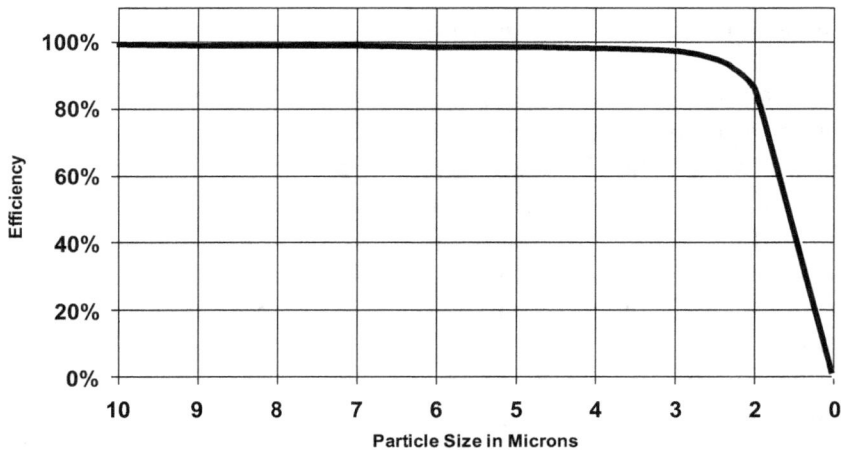

Figure 5-1: Typical Media Efficiency Curve

As you can see, the media has a high retention efficiency for particles that are 10 microns and larger in average diameter. As the particle size decreases, so does the efficiency. To capture smaller particles at a higher efficiency requires the use of a different media that has a higher particle retention for the smaller diameter contaminates. It should now be apparent that filter manufacturers will need to know not only what particle size is being generated, but also what efficiency is required so they can supply the correct media type for an application.

The next important filtration term is **face velocity**. Face velocity is simply the term used to describe the volume of air travelling through one square foot of filter media. To determine face velocity, divide the total ACFM that will be going through the filter element by the number of square feet of media in that filter element. This will result in a face velocity number rated in Feet per Minute. For example, if a 150 ACFM vacuum pump is pulling air through a filter element with 3 square feet of media, the face velocity is:

150 ACFM ÷ 3 Square Feet = 50 Feet/Minute.

Typically, the higher the face velocity, the lower the efficiency of the media. In other words, the faster the air has to travel through the element, the less efficient it will be at removing particles. Table 5-3 below highlights some industry guidelines for matching face velocity with general industrial vacuum applications.

Filtration Efficiency Requirements 99.7-% efficiency or better @	Environmental Conditions	Air to Media Ratio
Industrial Grade 10-micron **Paper**	Light Duty (clean, office/warehouse-like)	20 to 30 CFM: 1 ft^2
	Industrial Duty (workshop, factory-like)	10 to 15 CFM: 1 ft^2
	Severe Duty (Foundry, Construction-like)	5 to 10 CFM: 1 ft^2
Industrial Grade 10-micron **Polyester**	Light Duty (clean, office/warehouse-like)	40 to 50 CFM: 1 ft^2
	Industrial Duty (workshop, factory-like)	30 to 40 CFM: 1 ft^2
	Severe Duty (Foundry, Construction-like)	15 to 25 CFM: 1 ft^2
Industrial Grade 4-micron **Polyester**	Industrial Duty (workshop, factory-like)	15 to 25 CFM: 1 ft^2
	Severe Duty (Foundry, Construction-like)	10 to 15 CFM: 1 ft^2
Industrial Grade 1-micron **Polyester**	Severe Duty (Foundry, Construction-like)	5 to 10 CFM: 1 ft^2
Industrial Grade 0.3-micron **HEPA** Glass	Industrial Duty (workshop, factory-like)	5 to 7 CFM: 1 ft^2
	Severe Duty (workshop, factory-like)	3 to 5 CFM: 1 ft^2

Table 5-3: Industry Filtration Guidelines
Courtesy of Solberg Manufacturing, Inc.
Itasca, IL

The final term that should be mentioned is **pressure drop**. Anything put in line between the vacuum pump and the process will generate some amount of pressure drop. Inlet filters typically will have pressure drop values that are measured in inches of water column. To provide a rough idea of the magnitude of the inches of water scale, one inch of water column is equal to about $1/28^{th}$ of a PSI or stated another way, an inch of water is equal to 1.868 torr. The standard measurement of pressure drop through a filter/filter element assembly is when the element is new and clean. In this condition, pressure drop can range anywhere from less than 1″ H_2O to over 20″ H_2O at full flow conditions. The lower the pressure drop the better, but the actual maximum pressure drop requirement is dependent on the application. Typically, most properly sized vacuum inlet filters exhibit anywhere from 2″ H_2O to 8″ H_2O when clean and at full flow conditions. Filter element change out should be completed long before pressure drop effects process results.

One item to note is that vacuum inlet filters should always be sized for full flow at the nominal rating of the vacuum pump. The reason is that even though the actual mass flow in SCFM under vacuum conditions may be very low, pressure drop is effected not only by the available filter media surface area, but also by pipe connection size. Remember that under vacuum conditions, the mass flow will always be less than the volume flow, so sizing for SCFM flow at vacuum instead of full ACFM flow will result in pipe connection sizes that are too restrictive. Restrictive pipe diameters will create excessive pressure drop in the system. As a good rule of thumb, assume the volume flow in ACFM is equal to the mass flow in SCFM. That way, the filter will always be sized for maximum flow and minimum pressure drop.

It would be practical at this point to mention something about particle size distribution. While in some applications, assumptions can be made regarding expected particle sizes, there are many applications where there will be no information whatsoever regarding the expected particle sizes. In these cases, it is necessary to send a sample of the particulate out for analysis to determine what size particles are present. There are service companies available that can run tests on a sample of the particulate and then generate a report that outlines the particle size distribution for that particular sample. With this type of information it is possible to select the proper media needed to protect the vacuum pump. It must be noted, however, that in "normal" vacuum applications, a nominal 10 micron filter element is sufficient to protect most vacuum pumps. See Figure 5-2 below for an example of a particle size distribution report.

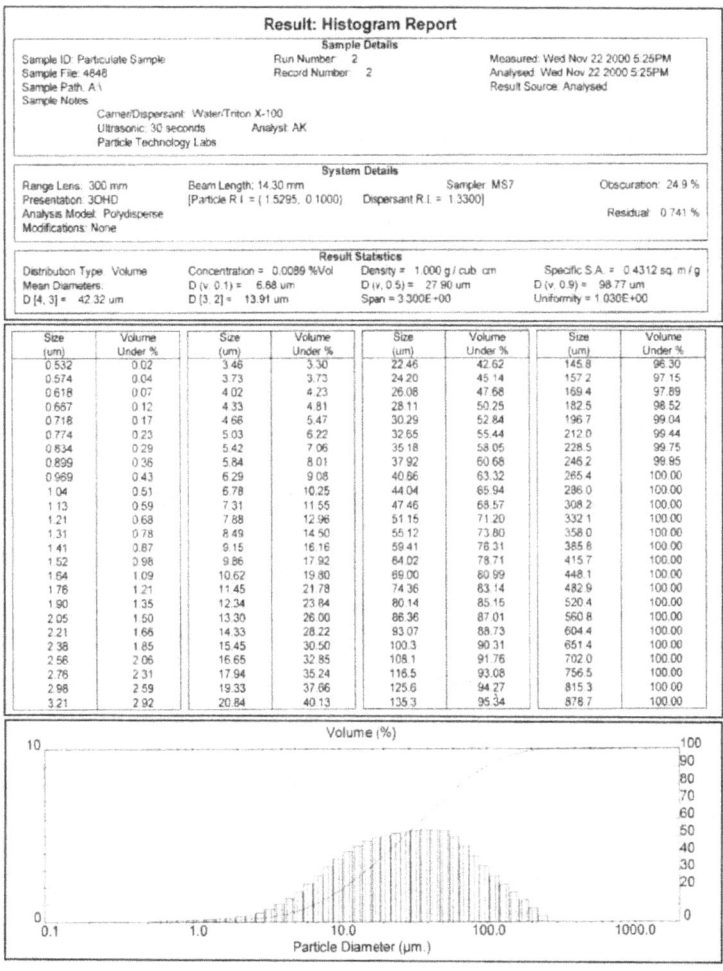

Figure 5-2: Particle Distribution Analysis

Courtesy of Particle Technology Labs, Downers Grove IL

One other question that should be addressed is the type of media that should be used. The two most common filter media types used in vacuum applications are paper and polyester. Polyester is used in a majority of vacuum applications because it is washable and works well in moist applications where there is the potential for liquids to be present in the vacuum system. Paper media, while it has better particle retention, can deteriorate quickly when used in wet applications. Both types work well and it is really up to end-users as to their preferences. There are many applications where paper and polyester media are not suitable. In these applications, there are other media types that should be considered.

Armed with the information on particle size, face velocity and pressure drop, the focus can now turn to the actual sizing of an inlet filter for a vacuum application. An example featuring a vacuum pump with a rated capacity of 500 ACFM in a hospital central vacuum system can be used to illustrate the inlet filter sizing technique. Knowing that hospital central vacuum systems typically generate very little particulate, it is possible to use the light duty face velocity criteria of 45 feet per minute (polyester). Also, standard 10 micron media elements are sufficient for most hospital applications.

For a 500 CFM vacuum pump in a hospital application, the first step is determining how many square feet of media are required. For light duty service (see Table 5-3), take the 500 ACFM pump capacity and divide it by the number of CFM per square foot of media in a light duty application. Remember that the ACFM and SCFM are assumed to be equal. The following equation results:

500 ACFM ÷ 45 Feet/Minute = 11.1 square feet of media.

Since there is the potential for light liquids to be present in this application, polyester media will be used. The next step will be to determine the amount of pressure drop that will be present when the element is placed in a housing and put in line. Most manufacturers have pressure drop curves available for each of their filter assemblies.

Typically, this information is readily available. Consult these curves to ensure that the expected clean pressure drop is not greater than the process can tolerate. Again, the average clean pressure drop is from 2" H_2O to 8" H_2O.

Connection size also has a lot to do with the expected filter assembly pressure drop. For a given vacuum flow in ACFM, a larger diameter connection size is better as far as pressure drop is concerned. In many cases, the connection diameter for the filter housing will be determined by the available connections on the vacuum pump. Note that these are usually adequately sized. There are, however, cases where the vacuum pump inlet connections are too small and specifications on piping connections called out by the user will result in excessive pressure drop. It is important to convey this information to the end-user so that proper decisions can be made regarding system and accessory piping.

In some instances, end users will give pressure drop specifications that are so low that just having the flange connections in line will create more pressure drop than is tolerable. In these applications, it is necessary to oversize the filter housing and use reducers to connect to the system piping. The larger filter system will be more expensive, but it is preferable to having the inlet filter be the source of excessive pressure drop.

Pipe Size (inch)	Area in Sq.Ft.	Velocity (ft/min)
0.5	0.001	366,692
1	0.005	91,673
1.5	0.012	40,744
2	0.022	22,918
2.5	0.034	14,668
3	0.049	10,186
3.5	0.067	7,484
4	0.087	5,730
5	0.136	3,667
6	0.196	2,546
8	0.349	1,432
10	0.545	917
12	0.785	637

Table 5-4: Velocities in a 500 ACFM Application

More will be discussed regarding pressure drop in vacuum piping later, but a brief discussion is necessary here so that pipe connections to filters can be understood in terms of velocity. **As a rule of thumb, the recommended velocity through a piping connection is approximately 92 feet per second (5,500 feet per minute) or less**. The way to determine velocity is to take the area of the pipe opening in square feet and divide the total ACFM flow by this number. For example, Table 5-4 illustrates the velocities present in the example 500 ACFM hospital vacuum system through various pipe connections.

You can see that the 5″ and larger diameter pipe connection velocities are below 5,500 feet per minute. Any one of these large diameter pipe connections are adequate for this particular example. Keep in mind that it does not typically create problems when using larger-than-recommended diameter pipe. Even though the volume of the pipe is greater, there is lower pressure drop across each component. The additional volume does not normally contribute enough to the capacity load to make a significant difference. For this example, pipe connections smaller than 4″ will generate higher-than-recommended pressure drop and could be the cause of process or end-use problems.

To summarize, the inlet filter for this application (500 ACFM vacuum pump in a hospital application) should be a polyester filter element rated for 10 microns. The element should have at least 11.1 square feet of media and the filter housing should have 5″ or larger diameter connections to the vacuum system. Note that this a light-duty application for a filter element. If the user wanted to oversize the element or the housing it would create no other problem than additional up-front cost. In fact, larger surface area filter elements typically provide for longer maintenance intervals.

Following is another example which will provide practice in the process of selecting and sizing an inlet filter. A woodworking shop has a 300 ACFM vacuum pump that supplies the vacuum used to hold down sheets of wood prior to CNC cutting and routing. The vacuum pump will need protection from a large amount of sawdust, wood chips and other debris. What is the correct size filter to place into this application? In most woodworking applications, 10 micron media is adequate, but it is still advisable to obtain a particle distribution, as sometimes smaller particulates are possible in applications that involve plastics or other composites. In this case, it is a severe duty installation and it is necessary to follow the guidelines for severe duty filter sizing (Table 5-3).

Severe duty requires an element with 15 to 25 ACFM per square foot of media. Dividing 300 ACFM by 20 (the average between 15 and 25), results in a requirement of 15 square feet of media for the filter element. To determine the proper connection size, divide 300 ACFM by 5,500 feet/minute which results in 0.05 square feet in pipe area. This corresponds to a 3.5" diameter connection (3" would also work). In some severe duty applications, you may find that the filter housing connection diameter needs to be larger than the supply pipe diameter so that filter elements with more square feet of media can be used. It is acceptable practice to use larger diameter pipe connections on the filter housing and reduce to the nominal system piping diameter. In this case, 3" (or larger) diameter connections are required on the filter housing along with a filter element that has 15 square feet of media. Nominal rating on the media should be at least 10 micron.

There is an alternative configuration for the inlet filter chosen for this particular application. Actually, there are a number of other alternatives, but only one will be discussed here. In many high particulate applications, it is common to use two filters in series to achieve the intended filtration goals. There are two reasons for doing this. The first is that the **lead filter**, or filter closest to the process, can be oversized to handle heavy particulate loading without short interval element change-out. The more surface area available on the filter element, the longer there will be low pressure drop across the element. This way, the process can run longer without having to clean or replace the filter element.

The second reason for using a lead filter is "operator" error. In many applications, machine operators are responsible for replacing or cleaning inlet filter elements without the assistance of qualified maintenance personnel. Many times in these situations, the filter elements are replaced without proper cleaning of element sealing surfaces. When this occurs, there is particulate bypass which results in contamination of the vacuum pump. With two filters in series, the downstream filter catches debris that bypasses the lead filter. This helps prevent contamination of the vacuum pump. When both filter elements are changed by unqualified personnel, however, the results can be the same.

In dual filter installations where the goal is to increase the maintenance interval, the lead filter element should have at least twice the surface area as the lag filter. Both housings should be installed in such a way so that when the elements are changed or cleaned, particulate does not fall into the system piping or into the vacuum pump itself. Some filter designs allow for element change-out and filter cleaning without the possibility of this problem occurring.

It is sometimes advantageous to have two filters in parallel instead of two filters in series. In this type of installation, there are valves on both filter inlet connections and both filter discharge connections. Operation consists of running process through one filter until significant loading occurs. At this point, the process air flow is switched over to the clean filter while the loaded filter element is cleaned or replaced. These types of installations are convenient when the vacuum system cannot be shut down for any reason and preventative maintenance must be done while the system is in operation.

One thing to note about filter elements is that as heavy loading occurs, the velocity of air through the remaining open media increases. This increase in velocity is the result of fewer passages for air to flow through and can actually decrease the efficiency of some types of filter media. Other possible problems with heavy loading are the build-up of differential pressure and potential for particulate bypass when the elements are changed. The best way to prevent these conditions is to change the element on a regular basis before excessive loading occurs.

Another good way to prevent these conditions is to add a device called a **prefilter** to the primary element. The prefilter is usually a foam 'sock' that completely covers the element and adds available surface area to the element without significantly adding to the pressure drop. Prefilters can add up to 50% more particulate loading capacity to an element and are removable and washable. When used in vacuum applications, they should be periodically removed and cleaned. Prefilters will add life expectancy to the primary filter element in most applications by keeping the heavy particulate load away from the element surface.

It may be necessary in some instances to find out how many square feet of media are in a currently installed element, especially if there are questions about upsizing the filter. The following steps will provide a very close approximation of the number of square feet of media in a filter element:

Step 1: Measure the **Pleat Depth** (it is usually close to the difference between element ID and OD) in inches.

Step 2: Count the number of **Pleats**.

Step 3: Multiply the number of pleats, times the pleat depth, times 2 (**Pleats × Pleat Depth × 2**) to get the **Length** measurement.

Step 4: Measure the element **Height**.

Step 5: Multiply: **Length × Height** to get the number of square inches then divide by **144** (number of square inches in a square foot).

This answer will provide you with the square feet of media in a filter element so that you can make decisions regarding how much larger a new element should be.

Keep in mind that inlet particulate filters are just that – inlet particulate filters. They are typically not designed to filter liquids or vapors which tend to pass directly through the media. The best way to remove liquids from the incoming gas stream is to use a different type of separator that will not only remove the liquids but will also provide a holding tank for the separated liquids. Chapter 7 has more information on these types of separators. Vapors are typically removed from the incoming gas stream with condensers or cold traps. In most cases they cannot be completely removed but only reduced to manageable levels.

For those applications where there are particles present that are smaller in diameter than 10 microns, there are other media available to assist in their removal. Standard media ratings for vacuum filter elements are 100 micron, 50 micron, 10 micron, 4 micron, 1 micron, 0.3 micron (HEPA) and 0.12 micron (ULPA). **HEPA**, or High Efficiency Particulate Arrestance, has a typical rating of 99.97% efficiency at 0.3 micron. **ULPA**, or Ultra Low Penetration Air, has a rating of 99.999% of all particles down to 0.12 micron.

Activated carbon granule elements or impregnated media wraps are used to help eliminate odors or take out small amounts of chemical vapors. Sizing for these types of filters is usually best left to filter manufacturers due to the variances in carbons and their subsequent holding capacities for adsorbed vapors. To illustrate, if the holding capacity of a particular carbon element is 15 pounds and the process generates a half a pound per day, then replacement or regeneration will have to occur every 30 days. This is far too short a time to be economically worthwhile. It is important to have adequate life expectancy from any process filter element, so good element design is critical.

There are also applications where mildly corrosive gases are present and particulate filtration is needed. Standard paper or polyester media can be attacked by the process gases in some of these cases. Standard element materials of construction may not be suitable for installation into those applications. There are other types of media available that can handle some of these corrosive gases. Materials such as **polypropylene, fiberglass and stainless steel mesh** can work well in these environments. As an additional note, many elements are made with steel endcaps and reinforcing wire mesh on the element outside diameter (OD) or inside diameter (ID). The endcaps and reinforcing wire can also be subject to corrosive attack. In these applications, special stainless steel or Teflon coated OD's,

ID's and endcaps are available to help prevent corrosion and chemical attack.

One other application deserves note: high temperature operation. There are many vacuum applications that involve high temperature inlet gas. Most standard filter elements have components which can fail when exposed to high temperature conditions. Typical temperature ratings for standard applications are 250° F and below. Higher temperature operation requires sturdy element construction and special media such as **Nomex** cloth. High temperature elements are built for applications where there can be prolonged exposure to temperatures as high as 400° F (and greater). Any vacuum pump can be protected from any portion of a process gas. The real question is how much is it worth to provide this protection. In standard applications, it is good practice to not only have adequate inlet filtration, but to keep it maintained at acceptable pressure drop and velocity levels. A good method of determining the effectiveness of a filtration program is to take samples of the vacuum pump seal/lubricating fluids on a periodic basis. A lubricant analysis will provide information on the condition of the lubricant in general and the level and type of particulate contamination. Figure 5-3 is an example of a typical particulate analysis report that highlights the level of particulate contamination for a vacuum pump. Note the breakdown by particle size.

PRODUCT ANALYSIS REPORT

CPI ENGINEERING SERVICES, INC.
2300 James Savage Rd. • Midland, MI 48642-6535
Phone: (517) 496-3780

Report Date:	01/15/2001
Report Number:	000285262

Customer	ABC Company
Comp. Mfr.	Vacuum Pump
Oil Type	Semi Synthetic
Serial Number	1234AZ
Model Number	AZ1111
Hrs. on Fluid	2500
Hrs. on Machine	2500
Sample Date	01/05/2001
I.D. #	000285262

ABC Company
123 Any Street

Any City, USA 12345
FaxNumber

Evaluation:

The fluid's parameters are normal. Sample again in 1000 hours.

Physical Properties* Results

Water by Karl Fischer (ppm)	Viscosity 40 C (cSt)	TAN Total Acid #	Particle Count								ISO Code	Antioxidant Level
			5 um	10 um	15 um	20 um	25 um	30 um	35 um	40 um		
24.80	98.39	0.124	9756.3	1546.2	409.2	266.7	217	193	172.6	150.9	20/16	
26.70	98.03	0.187	2419.9	1235.3	1110.9	1018.5	908.5	896.8	880.1	859.5	18/17	
14.80	101.3	0.129	1723.3	440.3	233.3	171.1	141.4	130.5	120.8	109.8	18/15	

* Property valuesshould not be construed as specifications

Spectrochemical Analysis

Sample Date (Lube Hours)	Wear Metals (ppm)										Contaminate/Additive Metals (ppm)							
	Silver (Ag)	Alum (Al)	Chrom (Cr)	Copp (Cu)	Iron (Fe)	Nickel (Ni)	Lead (Pb)	Tin (Sn)	Titan (Ti)	Vanad (V)	Bari (Ba)	Calc. (Ca)	Mag (Mg)	Mol. (Mo)	Sodi. (Na)	Phos. (P)	Sili. (Si)	Zinc (Zn)
01/05/2001 (2500)	0	0	0	0	0	0	0	0	0	0	0	0	0	0	0	0	0	0
12/21/2000 (1500)	0	0	0	0	0	0	0	0	0	0	0	0	0	0	0	0	0	0
12/05/2000 (500)	0	0	0	0	0	0	0	0	0	0	0	0	0	0	0	0	0	0

Thank you for this opportunity to provide technical assistance to your company. If you have any questions about this report, please contact us a 1-800-637-8628, or fax 1-517-496-2313.

CC List

* means this parameter not tested

ABC   496-0316

Figure 5-3: Lubricant Analysis Report
Courtesy of CPI Engineering - Midland, MI

DESCRIPTION OF TERMINOLOGY

Kinematic Viscosity @40 degrees C: ASTM D-445

Kinematic Viscosity is the property measured when a fixed amount of oil flows through a capillary tube under the force of gravity. The unit of Kinematic Viscosity is the stoke or centistoke.

Acid Number: ASTM D-974

Acid number is a number expressed in milligrams (mg) of potassium hydroxide needed to neutralize the acid in one gram of oil.

Antioxidant Level

High Performance Liquid Chromatography (HPLC) is used to determine the antioxidant level. Antioxidant levels of used fluids are compared to a top standard.

Metals Analysis

The Direct Current Plasma (DCP) runs metals analysis. Metals analysis detects both wear and additive metals. This instrument detects dissolved metals present @20 micron or less.

Water by Karl Fischer

The Karl Fischer conducts water analysis. Water content is reported in parts per million (PPM).

Particle Count

The Hiac Royco Particle Count indicates the number of particles present at different micron sizes. Both metallic and nonmetallic particles are counted. The ISO# is the ratio between the particles present @5 micron versus the particles present @ 15 micron. ISO# is reported like this:
15/12.

Chapter 6

Oil Mist Attenuation

There has been quite a bit of attention given to the discharge side of oil sealed vacuum pumps in recent years. The reason for this is the potential for oil carryover from vacuum pump discharge air to the surrounding ambient environment. With work areas becoming cleaner and regulations becoming more stringent regarding airborne oil contamination, there is a heightened awareness about vacuum pump discharge air. This awareness has resulted in an effort to not only keep the air clean and free of oil smoke, but also to keep production machinery, factory walls, ceilings and roofs all clear of oil contamination. Historically, oil sealed vacuum pump exhausts were vented to the outside of a building or vented directly in to the ambient air surrounding the vacuum system. In most areas, there are now governmental restrictions which prevent this. The bottom line is that oil sealed vacuum pumps can discharge oil even when supplied with good OEM separation packages. This chapter is dedicated to the attenuation of residual oil carryover from oil-sealed vacuum pumps.

As a starting point, this section provides an explanation of how an oil sealed vacuum pump works and what occurs in a typical OEM supplied separation system. Process air is drawn in from the vacuum system through the vacuum pump inlet and into the pumping module. Inside the pumping module, process air is mixed with vacuum pump seal or lubricating oil while compression back to atmosphere is taking place. The resultant air/oil mixture is then discharged from the pumping module into a separation tank where there may be one or several stages of separation prior to a final coalescing element. The heavy mist and large droplets are knocked out by the first few mechanical separation stages and the final oil "smoke" then proceeds to the coalescing element for final separation.

See Figure 6-1 for the layout of a pump oil circuit. Typically, there is a significant rise between the temperature of the inlet gas stream air and the temperature of the air/oil mixture on the discharge side of the vacuum pump. This is due to the heat of compression. Most OEM supplied separation systems do an adequate job of air/oil separation, bringing the oil carryover rate down to an industry standard 2 PPM (parts per million) to 5 PPM by weight or less.

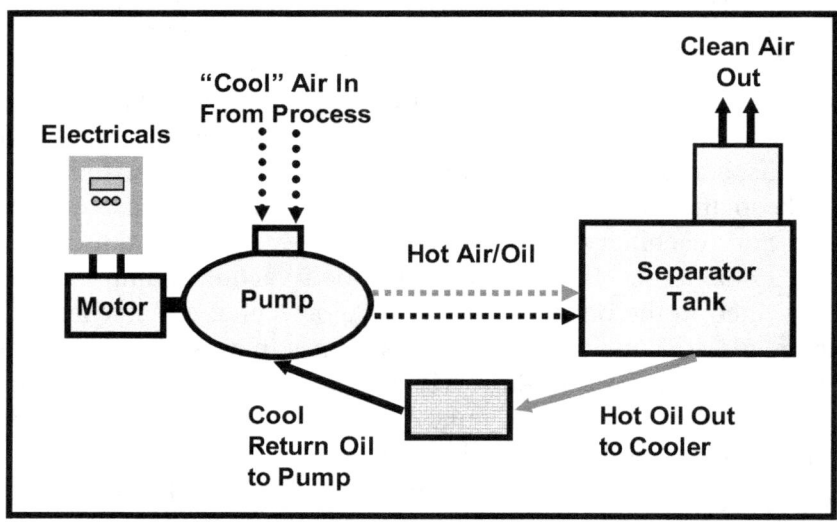

Figure 6-1: Oil Sealed Vacuum Pump Seal/Lubricant System

Similar to the chapter on inlet filtration, it is important to know industry oil separation terminology so you can talk the talk. Therefore, it is prudent at this point to take some time to learn and understand a few of the terms used to describe and quantify oil carryover. An understanding of these terms is helpful when discussing oil mist separation products and procedures.

Coalescing: This is the process by which small oil particles are combined into larger particles which are then separated from an air stream via some type of media. There are several physical processes that can occur depending upon the initial size and velocity of the oil particles. This term is most commonly used in conjunction with an oil mist separator element.

Back Pressure: The amount of pressure restriction generated by the separation process. Back pressure is typically measured in either PSIG or Inches of Water Column ("H$_2$O). Baffles, restrictions, changes in flow direction, separator elements and discharge piping can all add to the total back pressure in a separation system. Typical back pressures associated with vacuum pump discharge systems range anywhere from 0 PSIG to about 8 PSIG. Note that there are some vacuum pump separator designs that generate higher back pressures than this. These are usually designed and built for specific applications where the inlet vacuum and discharge pressure are both utilized for a specific process.

PPM: Parts per million by weight of oil (typically) and is a measure of oil carryover from a separation system. Industry standard carryover rates range from 2 PPM to 5 PPM by weight (weight of oil to the weight of air) but can be much higher in some applications.

Mg/M^3: Milligrams of oil per cubic meter of air. Another way to express oil carryover from a separation system. Conversion from Mg/M^3 to PPM is: Mg/M^3 × 0.83 = PPM.

Discharge Temperature: The temperature of the air/oil mixture as it is exiting the vacuum pumping module. Most oil sealed vacuum pumps operate in the discharge temperature range from 140° F to 240° F with about 170° F being the most common.

Pre-separation: The mechanical separation of heavy oil droplets and mist prior to reaching the coalescing element.

Draining: When a separator element reaches the saturation point, the coalesced oil will begin to gravity drain. If separation is done from the inside of the element out, much of the oil will drain back down into the oil reservoir. An additional amount of oil will be forced through the separator element which then gravity drains on the outside wall of the separator element. This coalesced/drained oil must be collected and scavenged back to the inlet of the vacuum pump. Otherwise, oil will build up to levels that eventually reduce the amount of surface area available for coalescing. In these situations, the subsequently higher velocity air through the coalescing element will carry more oil with it in the exhaust gas stream.

For outside-in separation systems, much of the coalesced oil will drain back to the reservoir from the outside of the element. The same scavenge arrangement is typically necessary for oil that is collected on the inside of the separator element (also called the "dry" side). See Figure 6-2.

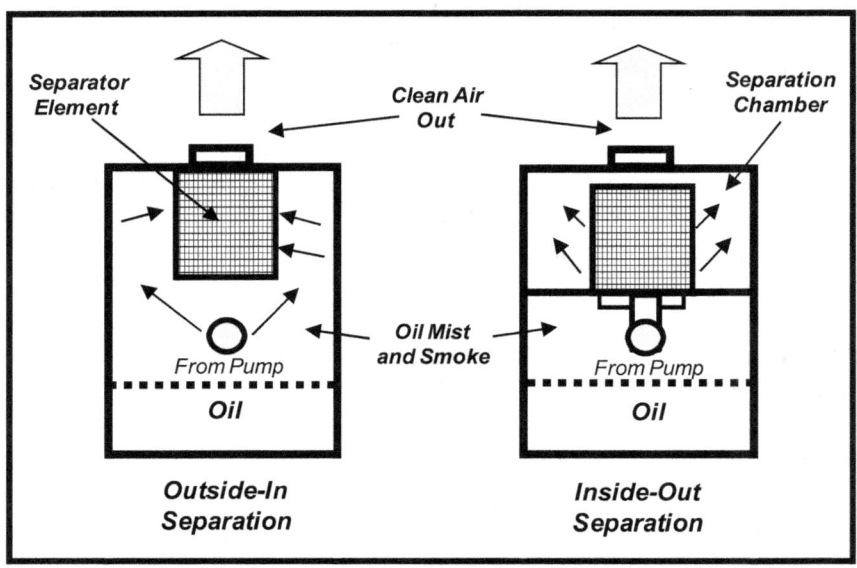

Figure 6-2: Inside-Out and Outside-In Separation

Discharge oil separation for vacuum pumps can be very tricky. Since inlet conditions can change from atmospheric pressure to full vacuum within a number of seconds, the separator element must be able to handle a full range of air flows and velocities. If the separator element is sized incorrectly (too small), there can be higher oil carryover rates. Temperature also plays an important role in the effectiveness of a separation system. Higher temperatures typically mean higher oil carryover rates with standard vacuum pump oils. The reason is that many vacuum pump oils turn into vapor at a faster rate when exposed to high temperatures. Oil vapors are very small particles and will go right through a separator element along with the discharge air without being coalesced.

The obvious solution would be to run the vacuum pump at lower discharge temperatures. The problem with lower discharge temperatures is the fact that when a vacuum pump operates with discharge temperatures that are too low, water vapor can condense in the oil reservoir. There are many applications where vacuum pump oil reservoirs have filled up with liquid water resulting from the condensation of inlet air stream vapor. More will be said on this in chapter 9 which deals with water and water vapor in vacuum systems.

In the average installation, a typical oil sealed vacuum pump will have an oil carryover rate of approximately 3 PPM by weight. Even though there is oil present in the discharge air stream, it is a seemingly small amount when compared to the total flow rate (3 PPM does not *seem* like a large quantity of oil). If that is the case, why are so many end-users asking for additional oil mist attenuation for their vacuum systems? The answer lies in the fact that for every 1 PPM of oil carryover, there is approximately 0.72 ounces of oil present per 100 hours of operation and 100 SCFM discharged from the vacuum pump. A carryover rate of 3 PPM would be: 3×0.72 which is a total of 2.16 ounces of oil for every 100 SCFM and 100 hours of operation.

To illustrate the magnitude of oil carryover at that rate, follow along with the next example. Take a 200 ACFM vacuum pump operating at 15″ HgV (to make the math easy on ourselves). At 15″ HgV, this vacuum pump continuously discharges 100 SCFM of air along with the industry standard 2 PPM to 5 PPM of oil. To illustrate how much oil is being carried over past the OEM separation system, review Table 6-1. The information in the table is based on the assumption that the average density of oil is 7.5 lbs/gal.

Hours	Total (ounces) Oil Carryover	2 PPM Fluid Volume	5 PPM Fluid Volume	Time Line
100	2.2	0.1 quarts	0.18 quarts	4.17 Days
500	10.8	0.4 quarts	0.9 quarts	3.0 Weeks
1000	21.6	0.7 quarts	1.8 quarts	1.4 Months
2000	43.2	1.4 quarts	3.6 quarts	2.7 Months
4000	86.4	2.9 quarts	7.2 quarts	5.5 Months
8760	189.2	1.6 gallons	3.942 gallons	1.0 Year

Table 6-1: Oil Carryover Rate for a 200 ACFM Vacuum System

From the table, it is shown that oil discharged at a 5 PPM rate totals about one quart every 3 weeks. If this oil is discharged into the ambient room or inside a sound enclosure around the vacuum pump, the oil will condense on the first cool surface it sees. Over the course of a year, the nearly 4 gallons of oil discharged can create quite a mess in an equipment room or in the actual work area.

In some cases, discharge air from vacuum pumps is piped through the factory roof or through an outside wall where it is expelled from the building. Over time, oil will accumulate on the exterior wall or on the factory roof. This condensed oil sometimes drips down the side of the building causing an unsightly mess. Therefore, the reason why many end-users are looking for additional oil mist attenuation is that, in addition to the fact that they do not want their employees breathing oil vapors, there is still a significant amount of oil that gets past standard OEM separation systems.

Now that magnitude of the problem is known, the next logical questions are: one, what are the options for additional mist attenuation; and two, how do you properly size a "polishing filter" as they are called for these applications? Within a reasonable price range, the only viable option is to put a canister type exhaust filter in the discharge line after the OEM supplied separation system. Many companies with large numbers of vacuum pumps in operation vent their vacuum pump discharge lines into a common scrubber system. These scrubber systems take out a high percentage of the oil mist from the air stream prior to discharge from the building. However, for systems with one or two vacuum pumps, this can be an expensive alternative.

Canister type filter housings with replaceable elements are a much more economical method for cleaning up vacuum pump discharge air. Not only is there the benefit of simple installation, the replacement filter elements are easy to change and cost effective if maintained properly. Oil mist filter elements that are not exposed to overly high temperatures, contaminated oil or significant amounts of particulate matter can last for years in most applications.

The sizing of oil mist exhaust filters is based mainly on nominal pump rating and economics. Normally, the process of sizing an oil mist exhaust filter is to match the nominal vacuum pump rating (rating at atmospheric pressure) with the SCFM rating on the exhaust filter. Over-sizing the filter is helpful because the larger available surface area on the filter element prolongs the period of low oil carryover. The polishing process starts with entrapment of oil particles on the surface and within the separator media and continues through saturation and draining. While the element is becoming saturated, residual oil carryover rates past a polishing filter is low at the beginning and begins to rise after complete saturation occurs.

Therefore, the best polishing filter performance occurs while the element is in the saturation phase. The more surface area available during the saturation phase, the longer the discharge air will contain low oil carryover. In general, more element surface area is better than less and from a carryover perspective, there is typically not an exhaust filter that is too large for an application. Normal carryover attenuation with a polishing filter is anywhere from 10 to 100 times when compared to the OEM supplied system alone. This puts the average carryover rate past a polishing filter at 0.3 PPM to 0.03 PPM. Very low carryover is considered to be 100 times lower than the carryover rate from the OEM system alone.

The economics side of the equation comes into play when users do not want to size for the nominal flow rating of the vacuum pump. The main reason for sizing an exhaust filter for lower than the nominal vacuum pump capacity is because a particular application will be operating continuously at a specific vacuum level. The user does not want to buy filters that are sized for nominal pump capacity when they can get away with sizing the filter for target capacity.

It is acceptable to size an exhaust filter for these conditions, but there is some risk associated with it. If the vacuum pump is inadvertently operated at higher pressures due to a leak or system upset, there is the risk of building excessive back pressure at the vacuum pump discharge. This is due to the higher mass flow through the vacuum pump during these conditions and the exhaust filter would not be sized to handle the higher flow. To illustrate the mass flows associated with various vacuum levels, refer to Table 6-2.

Nominal Pump Rating: 150 ACFM	
Vacuum ("HgV)	SCFM Flow
0	150
5	125
10	100
15	75
20	50
25	25
28	10
29	5
29.5	2
29.9	0

Table 6-2: Mass Flow for a 150 ACFM Vacuum Pump

As is shown, operation at 20″ HgV yields a discharge air flow of approximately 50 SCFM on a vacuum pump nominally rated for 150 CFM. When the higher discharge temperature of approximately $170°$ F is accounted for, there is actually 59 ACFM at the discharge port of the vacuum pump (remember Charles' Law). This is the true target capacity to aim for when adding a polishing filter. In most cases, however, accounting for discharge temperature will not make a huge difference in sizing. What it does is provide for a safety factor when choosing between two different size filters.

So, given a choice between a filter rated for 150 SCFM or a filter rated for 59 SCFM, the obvious economic choice is the filter rated for 59 SCFM. There is a word of caution, however. Filters sized for 59 SCFM will have approximately a 1 ½″ inlet and discharge connection whereas a vacuum pump rated for 150 CFM will have inlet and discharge connections of 2″ or 2 ½″. Worst case scenarios usually do not occur, but excessive back pressure could result from full flow conditions. When exhaust filter sizing is based on economic considerations, it is best to build in a buffer, 20% for example, to account for variations in process flow or the development of a significant leak. It is imperative that complete information is known about the process so that the oil mist exhaust filter can be properly sized.

The next drawing illustrates the typical setup for an installation of a polishing filter on a vacuum pump that has an OEM supplied primary separation system. Please refer to Figure 6-3. One question that arises after reviewing this arrangement is what should be done with the coalesced oil after it collects in the polishing filter? In most cases, it is drawn back to the inlet of the vacuum pump through a scavenge tube that utilizes inlet vacuum to transport the oil.

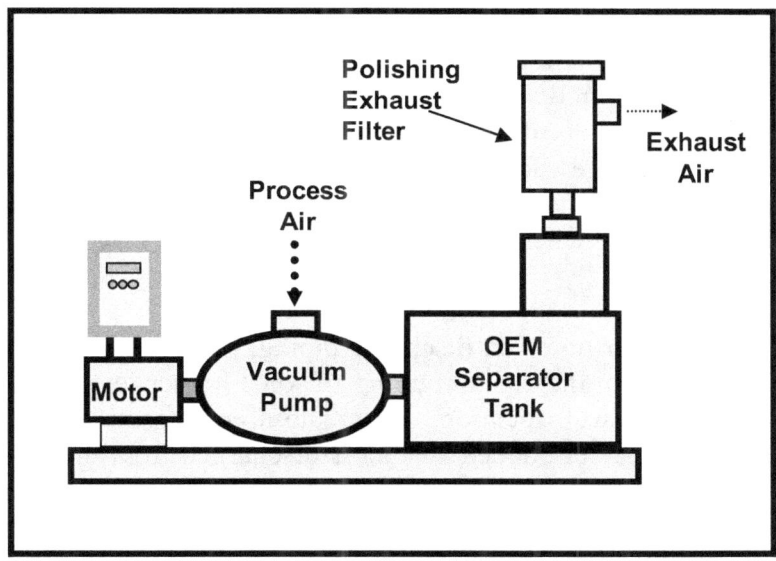

Figure 6-3: Typical Polishing Filter Arrangement

If the vacuum level is sufficient, the suction line back to the inlet (the **scavenge line**) can remove the coalesced oil as it is collected in the bottom of the separator housing. If the inlet vacuum is insufficient or if there is debris in the scavenge line, the oil will not be removed rapidly enough and it will begin to build up in the separator housing. If the oil level in the separator housing becomes too high, there will be less element surface area available for the discharge air to flow through. Higher back pressures and higher air velocities will be the result. Both of these conditions will reduce the efficiency of the polishing filter. Other than draining the coalesced oil back to the inlet of the vacuum pump, it is common practice to gravity drain it directly into a proper storage container. This oil can then be filtered and reused or properly disposed.

One condition where it would not be advisable to utilize a scavenge line is where there is the possibility for water or other contamination to collect in the oil mist separator housing. For example, if the process generates a significant amount of water or water vapor, there is the potential for condensed water to collect in the separator housing. If this condensed water mixes with the coalesced oil, they will both be drawn back to the inlet of the vacuum pump. This condition can create a continuous water loop where the water is endlessly recycled and builds up to problematic levels. Depending on the type of pump being used, this can have a serious effect on the mechanical operation of the vacuum pump or on the pump's ability to draw an adequate vacuum for process. The same can be said for processes that generate other types of vapors as well. These vapors can condense anywhere in the discharge line or exhaust filter housing.

A couple of other notes on discharge piping. Discharge piping that rises directly from the vacuum pump to some higher level, be it the ceiling or upper wall location, can create a condensation problem. As the warm, moisture laden air that is discharged from the vacuum pump rises through exhaust piping, it cools. As it cools, the water vapor entrained in the air stream can condense in the piping and drip back down into the vacuum pump, separation system or polishing filter. This situation can also result in a water loop that contaminates the vacuum pump.

The proper way to discharge vacuum pump exhaust is to use a drip leg with a drain valve. See Figure 6-4 for both the proper and improper methods to install discharge piping on a vacuum pump. Another note about discharge piping is to be sure to use proper diameter pipe. More will be said regarding the proper diameter of vacuum piping in Chapter 7, but suffice it to note here that the discharge piping should not create any more than about 0.5 PSIG total back pressure at full flow.

High back pressure can hinder the vacuum pump's ability to pull adequate vacuum and will definitely cause an increase in operating horsepower. This holds true not only for single pump installations, but for multiple pump installations as well, where discharge lines from several vacuum pumps are tied together into a common manifold.

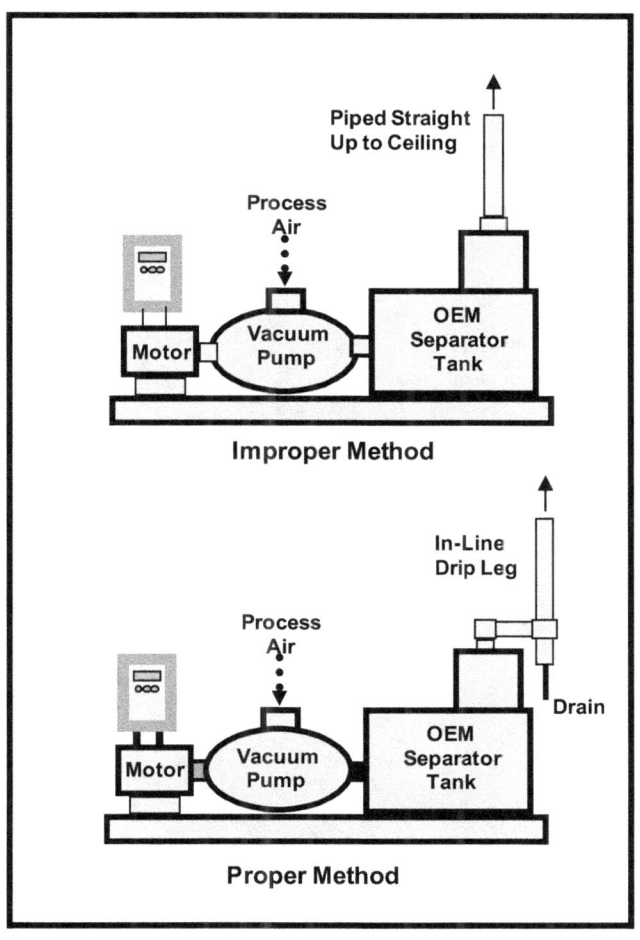

Figure 6-4: Recommended Discharge Piping Arrangement

A note for vacuum pump packagers: there are alternatives to developing your own air/oil separation system and taking chances on attaining good air/oil separation. There are pre-designed and pre-packaged vacuum pump separator systems available that can be customized to any type of vacuum pump. These systems provide an oil reservoir, do an excellent job of liquid pre-separation and are supplied with coalescing elements for final attenuation of fine mist and smoke. One of the keys to good oil mist attenuation is good pre-separation and these systems do an excellent job. See Appendix B for manufacturers of this type of equipment.

Chapter 7

Vacuum Piping and Pressure Drop

In any vacuum system, there exists the potential for measurable and potentially costly pressure drop. In this chapter, the causes of excessive pressure drop will be examined along with how to prevent it from creating excessive economic cost. Following that are discussions about vacuum accessories, vacuum gauges and how to properly size piping for vacuum applications.

Users typically take great care in the selection of process vacuum pumps. Vacuum pumps are evaluated and selected based on factors such as performance, energy utilization, size, sound level and price. The components that are installed in-line between process vacuum pumps and production machinery are vitally important to the proper functioning of the system. Having said that, the distribution system is usually one of the most neglected aspects of vacuum supply. It is important to properly design the vacuum piping and distribution system because of the problems that can result from excessive pressure drop.

When excessive pressure drop exists in a system, vacuum pumps must be operated at elevated vacuum levels to compensate for the high pressure drop. Operation at higher vacuum costs energy dollars and magnifies leak problems. Simple and cost effective measures can be taken to alleviate the problems associated with excessive pressure drop in vacuum systems. These measures will raise the operating efficiency of the system and provide for an increase in the available vacuum supply to production.

What is pressure drop when discussed in relationship to a vacuum system? Pressure drop in a vacuum system is similar in concept to pressure drop in plant compressed air systems. It is simply the difference in operating pressure that exists between the supply point and the use point. In compressed air systems, pressure drop is typically measured in PSI, which, as you recall, is a common pressure scale that stands for pounds per square inch. PSI can be used as a measure of pressure drop in vacuum systems as well, but it is much more common to use inches of mercury ("Hg). Note that the "V" and "A" are missing from "Hg. Pressure drop is a measure of the *difference* in pressure, so whether it is inches of mercury vacuum or inches of mercury absolute is irrelevant. You learned in Chapter 1 about the various scales used to measure vacuum.

Again, refer to Table 7-1 which illustrates the most common vacuum scales and how these scales compare with one another. Remember that the top of the table represents atmospheric pressure and the bottom of the table represents perfect vacuum. Pressure drop can be measured in any one of the units listed below.

PSIA	"HgV	"HgA	Torr	Mbar
14.7	0	29.92	760	1,013
13.7	2	27.92	709	946
12.2	5	24.92	633	844
9.8	10	19.92	506	675
7.3	15	14.92	379	505
5.9	18	11.92	303	404
4.9	20	9.92	252	336
3.9	22	7.92	201	268
2.4	25	4.92	125	167
0.9	28	1.92	49	65
0.5	29	0.92	23	31
0	29.92	0	0	0

Table 7-1: Common Vacuum Units

In a compressed air system, the typical system pressure drop can be as high as five or ten PSI. You can see from the above table that five PSI is equal to about 10" Hg. In a vacuum application, a normal pressure drop of 10" Hg would be an example of a very poorly functioning system. The result is that the total pressure drop in a vacuum system has to be much lower than in a comparable compressed air system.

There is a simple and cost effective method can be used to find the total amount of pressure drop in a vacuum system. First, measure the vacuum level at the inlet of the vacuum pump using a reasonably accurate vacuum gauge. Then, with the same gauge, measure the vacuum level at the point of use or as close to the point of use as possible. Note that it is acceptable to use two separate gauges, but calibrated gauges or gauges that display similar values for the levels of vacuum found in the system should be used instead of off the shelf gauges. The difference between the vacuum level at the pump inlet and the vacuum level at the end-use point is the total system pressure drop.

Depending on the type of system, the product manufactured and the operating vacuum level, average system pressure drops in vacuum applications will vary from a fraction of an inch of mercury up to 15″ Hg. Higher pressure drop equates directly to higher operating costs.

To illustrate how much artificial demand is created by pressure drop in a vacuum system, an example of a process that requires a volume flow of 100 ACFM at 20″ HgV will be used. This example (100 ACFM) is being utilized because it can be scaled up or down to fit many different applications. As pressure drop increases, there is a need for higher vacuum at the pumping system to compensate for losses in the system. Although this may not be a problem in moderate levels, higher vacuum and higher capacity systems can suffer significantly from increased pressure drop. A relatively large vacuum pump installed in a low demand application may actually have extra capacity available (note that some vacuum pump technologies actually use less brake horsepower as vacuum increases) so pressure drop will have little effect on the cost of running that system.

In the more severe cases, however, extra pumping capacity will have to be added to the system to account for the greater volume of air. The additional capacity is required because in general, as vacuum level increases, air entering the system expands in proportion to the vacuum level (this is Boyle's Law from Chapter 2). Higher vacuum equals greater expansion. To attain the desired vacuum at the process use point and overcome the expansion of air, more volume capacity is needed. At some point, the installed vacuum pump will not be able to keep up with the expansion of air.

The amount of additional capacity required will depend on the operating vacuum level and the amount of pressure drop in the system. Table 7-2 illustrates how this phenomenon affects our example system.

ACFM Required By Production Process @20" HgV	Delta P In Dist. System	Required Vacuum At Pump	ACFM Required To Compensate For Pressure Drop
100	1" Hg	21" HgV	111
100	2" Hg	22" HgV	125
100	3" Hg	23" HgV	142
100	4" Hg	24" HgV	166
100	5" Hg	25" HgV	200
100	6" Hg	26" HgV	250
100	7" Hg	27" HgV	333

Table 7-2: Additional ACFM to Overcome Pressure Drop

Table 7-2 shows how pressure drop can add significantly to the number of vacuum pumps required to run a production process. For example, in this system, a pressure drop of 3" HgV adds 42% to the required ACFM flow. It is not difficult to see how reductions in system pressure drop can lower energy and operating costs by reducing the number of on-line vacuum pumps. Operating fewer vacuum pumps not only saves energy dollars, it also lowers overall maintenance costs. Other non-tangible benefits to fewer on-line pumps include lower noise level, less oil mist carryover and lower ambient heat loads. Finally, if off-line pumps can be eliminated, usable floor space can also be increased.

Operation of a vacuum system at higher vacuum also affects the volume flow of air entering the system through leaks. Air entering a vacuum system through leaks adds to the production demand load and must be treated as if it were production demand. As the vacuum level increases, the effect of leaks on the system increases as well. If the example system has 6 ACFM in leak flow rate at 20" HgV, then operation of the same system at 25" HgV will double that number. Even though the percentage of total flow remains the same, it is still an additional load on the vacuum pumps. Also, depending upon the design characteristics of the distribution system, operation at higher vacuum levels may open more leaks due to the increased differential pressure. Chapter 9 covers leaks and leak detection.

Once total system pressure drop has been measured, it is necessary to determine which components are adding the most restriction. It is recommended that each component be reviewed individually and then ranked against all other components so a "worst first" repair program can be implemented. To check an individual component, tap into the inlet and discharge of the component and measure the pressure drop. Pressure drops for each system component from the vacuum pump to the point of use should be measured and recorded. Items like inlet filters with replaceable elements should have running logbooks established so that element change-out intervals can be determined. The following list contains some areas that should be checked or reviewed for each in-house vacuum system.

Piping: The single biggest problem with vacuum system piping is inadequate diameter (more on this later). The combination of restrictive pipe diameter and lengthy piping runs can create significant pressure drop. As a rule of thumb on single vacuum pump applications, maintain the diameter of the vacuum pump inlet as far into the process as possible. Smooth interior walled pipe is superior to rough walled. Piping should be kept short and straight. Elbows should be kept to a minimum and, where they are necessary, large radius elbows are preferable to straight 90-degree turns. On multiple vacuum pump applications, a full analysis should be completed to determine the optimum pipe diameter.

Valves: Isolation valves and check valves should be inspected to ensure they are full port and match the diameter of the system piping. Typically, standard ball valves have port diameters that are restrictive for vacuum applications. Full port ball, gate or butterfly valves provide excellent flow characteristics and very little restriction. Check valves can also be a source of restriction in vacuum piping systems. When check valves become lodged in a partially closed position or fail to completely open, immediate repair or replacement is required.

Filters: As discussed earlier, many vacuum pump technologies require inlet filtration to remove particulate from the incoming air stream. Filter element loading increases pressure loss and can be easily avoided with proper preventative maintenance. Improperly sized filters with small port diameters can also be a major source of restriction. Check with the filter manufacturer to ensure proper sizing and installation.

Receivers/Separators: At times it is necessary to remove liquids from the vacuum air stream prior to the vacuum pump inlet. It is important to have the correct type, configuration and porting on receivers and separators to ensure adequate liquid separation and low pressure drop. Many liquid separators have minimum and maximum velocity requirements for optimum separation efficiency. It is important to follow these guidelines so that maximum protection is provided for the vacuum pump.

Production Machinery: Production machinery sometimes accounts for the majority of system pressure drop. Conventional thinking, however, does not allow for changes to the internal plumbing of production machinery. Given that the thought process for production machinery design usually does not take into account the energy usage of vacuum supply, it is worthwhile to look into what changes can be made that will improve flow and not sacrifice production efficiency. Sometimes, improvements can be as simple as enlarging the internal diameter of supply tubing.

Vacuum Pump Controls: Some vacuum pump technologies utilize control mechanisms on the pump inlet to automatically regulate the system vacuum level within a preset range. These mechanisms are sometimes set incorrectly or are out of adjustment. Improper functioning of vacuum pump controls can choke off the airflow to the pump and appear to be plumbing problems even if the rest of the system is functioning optimally. Only qualified service personnel should adjust vacuum pump inlet controls.

Leaks: No vacuum system evaluation is complete without a leak check. Leak checks are important in some facilities because considerable horsepower is used just to overcome the system leak rate. There are several different techniques commonly used for detecting leaks in vacuum systems which are discussed in Chapter 9. Two very common methods are ultrasonic detection and tracer gas detection. Both methods are suitable for production vacuum systems.

In many cases, vacuum distribution problems are solved by adding bigger vacuum pumps to overcome system pressure drop. A program that identifies and corrects the significant causes of pressure drop has the potential to forestall or completely eliminate the need for new or additional vacuum equipment. Of course, it is not practical or economically feasible to eliminate all pressure drop from a vacuum system, but it is possible to eliminate pressure drop from the worst components so that the trade-off between operating costs and costs for changes in the distribution system is favorable. A proactive program can assist in taking vacuum pump horsepower off line. As an example of the amount of savings that can be realized for reducing operating horsepower, one 40 horsepower vacuum pump taken off line can result in a yearly savings of $16,845 at $.06/KWH and 8,750 operating hours. This is a significant sum, considering the typically nominal investment in time and plumbing changes.

The other advantage of a proactive approach is an increase in the quality of vacuum supply to end-use points. Once evaluation and repair programs are completed, vacuum distribution systems that have had increasing demands placed upon them over time or that were marginally sized to begin with, will not be as susceptible to fluctuations in production vacuum load. This results in more production up time, faster cycles, better-formed products and increased holding force. In other words, the system will have greater efficiency. Resist the tendency to add horsepower to solve vacuum supply problems. Before purchasing additional vacuum equipment or adding on-line horsepower to solve production vacuum problems, evaluate the vacuum distribution system for excessive pressure drop. It is an effective approach for both cost reduction and cost avoidance.

VACUUM PIPING
It is now time to talk a little more definitively on vacuum piping. There are essentially three primary issues that must be resolved in any vacuum piping installation. These three issues are pipe diameter, piping configuration and use of the correct materials of construction. In addition to these three issues, considerations regarding the selection and placement of vacuum accessories should also be addressed.

The most common questions regarding vacuum system piping relate to pipe diameter with the most obvious question being: what is the correct diameter pipe to use in a vacuum pump application? The answer to this question can become quite involved and is based on a number of factors which include: operating vacuum level; distance of the vacuum pump to the process; gas stream temperature; capacity rating of the vacuum pump and the type of gas being pumped from the system. To simplify things, why not just use the next larger diameter pipe size over the inlet connection to the vacuum pump? This solution would seem to provide low pressure drop and full process flow. The reasons we avoid this solution are twofold.

First, economics dictate that the most cost effective solution should be provided to connect the vacuum pump to the process. Even though a 12″ diameter pipe on a 300 ACFM system would provide for fractional pressure drops, the costs to install it would be tremendous. Second, fitting the physical piping system in the plant is a problem in many facilities. In many central vacuum systems, the piping system is run throughout the ceiling and must be threaded in and around support structures, electrical conduit, water and compressed air piping. This makes it at best difficult, or at worst impossible to install large diameter pipe. It is therefore important to size the vacuum system piping diameter as close to ideal as possible.

Having said that, please keep one thing in mind when evaluating a central vacuum system supply layout. In most applications, the vacuum supply system is designed and installed to meet current production machinery requirements. Over time, as production equipment is added or moved, piping and other accessories are added to the distribution system to accommodate these equipment changes. Eventually, the supply system looks like a mass of tangled vines and has lost its original function and structure. It is wise to design a vacuum supply piping system with a little bit of extra thought into what would happen if the plant is expanded or changed in some manner. The cost for slightly modifying the about-to-be installed system today to accommodate for tomorrow's growth is minimal compared to the cost for massive system changes later.

The rough sizing for vacuum pipe diameters can be done fairly quickly. Please note that there are calculations that can be performed to evaluate any circumstance: flow, vacuum level and different gases. These calculations are beyond the scope of this text. Here we will focus on general rules that can keep you out of trouble. **As a simple answer to the question of pipe diameter, in a one pump system, maintain the inlet diameter of the vacuum pump as far into the process as possible and keep the piping as short and as straight as possible**.

Vacuum pump manufacturers typically design their vacuum pumps with inlet port diameters that are adequate for this rule of thumb. Sizing pipe diameters based on this rule of thumb will result in a satisfactory level of system pressure drop.

In multiple pump systems, the situation is less clear, since total flow from all the vacuum pumps must be taken into consideration. If multiple pump systems were sized based on the rule of thumb above, there would certainly be higher pressure drops than are tolerable. As an example of the amount of pressure drop that can be present in a typical installation, consult Table 7-3. Table 7-3 outlines pressure drop across pipe of various diameters and 100 foot lengths at 20″ HgV and 150 ACFM. This is shown only to illustrate the differences in pressure drop from an expanded flow rate through increasing diameter pipe.

Pipe Size (inch)	Pressure Drop
1.5	50 torr (2″ Hg)
2	26 torr (1″ Hg)
2.5	8.2 torr (0.3″ Hg)
3	3.2 torr (0.1″ Hg)
3.5	1.4 torr (0.06″ Hg)
4	0.7 torr (.03″ Hg)
5	0.2 torr (0.009″ Hg)
6	0.08 torr (0.003″ Hg)

Table 7-3: Typical Pressure Drop in a Vacuum System

Looking at the chart, it is plain to see how important it is to have the correct diameter pipe. For example, note the difference in pressure loss for 150 ACFM through 1.5″, 2″ and 3″ diameter pipe. Also note that trying to pull too much ACFM through a given pipe diameter will only lead to additional pressure drop problems. For an incremental investment in larger diameter piping, there is a significant savings in operating dollars as well as a pickup in the available vacuum level at the point of use.

From a general design perspective, a rule of thumb pipe sizing criteria can be established based on the velocity of air through a pipe. Although this will not be accurate in every case, it can be generally applied in many situations to see if the projected diameter pipe is in the right ball park. It is important to establish a target maximum pressure drop for the entire system during the design of a vacuum piping system. Process needs will dictate the fine tuning of the piping system, but a general design criteria of 1″ Hg maximum pressure drop is a good rule to follow. This means that for most vacuum systems, strive to keep the pressure drop from the vacuum pump to the point of use at 1″ of Hg or less.

This rule applies to the sum of all system components which includes accessories such as filters, check valves, isolation valves and receivers that are in line between the vacuum pump and process. This is a general guideline only. In medium and high vacuum systems, a rule of thumb that establishes a 1″ HgV maximum pressure drop would be absurd! But for many rough vacuum systems, it provides a good cost/benefit ratio between the economics of installing an ideal piping system and installing a realistic piping system that works well for its intended purpose.

As mentioned earlier, the basis for establishing pipe diameters is the velocity of air through a pipe opening. The goal is to maintain a velocity of approximately 6,000 feet per minute (100 feet/sec) through any given portion of the piping system. To determine the air velocity, take the cross sectional area of the pipe in square feet and divide the air flow in ACFM by this number. Based on the amount of ACFM flowing through a given diameter pipe, results from real life systems are anywhere from a few hundred feet per minute to over 30,000 feet per minute.

Table 7-4 provides the approximate cross sectional areas (in square feet) of many common pipe sizes.

Diameter	Area (sq.ft.)
1/4"	0.0003
1/2"	0.001
3/4"	0.003
1"	0.005
1.5"	0.012
2"	0.022
3"	0.049
3.5"	0.067
4"	0.09
5"	0.14
6"	0.2
8"	0.35
10"	0.55
12"	0.79

Table 7-4: Cross Sectional Areas of Pipe in Square Feet

To establish the air velocity for a proposed or installed system, first find values for the capacity of the vacuum pumping system(s) and the diameter of the pipe that connects the vacuum pump to the process. For example, if a 250 ACFM vacuum pump is installed in an application with a 1″ inside diameter (ID) inlet pipe connecting it to a manufacturing process, the velocity of air is calculated by dividing the 250 ACFM by 0.005 square feet (surface area of the 1″ diameter pipe). The answer is 50,000 feet per minute.

This is an extreme example, but it illustrates that for this particular application, the pressure drop is going to be excessive and vacuum pump capacity will be wasted. To determine the actual pressure drop, it is necessary to know additional information about pipe length, vacuum level and temperature. For this application, what diameter pipe would be appropriate? As a first step in determining the correct pipe diameter, a table can be developed that compares the velocities resulting from using each of several standard pipe sizes. The ACFM flow rate from the pumping system can be changed for what-if type scenarios. The results are then reviewed and a determination is made on pipe size. See Table 7-5 for the results of a 250 ACFM application.

ACFM Flow	Pipe Size	Velocity (Ft./min)
250	1.5" / 0.012 sq. ft.	20,833
250	2" / 0.022 sq. ft.	11,363
250	3" / 0.049 sq. ft.	5,102
250	4" / 0.067 sq. ft.	3,731
250	5" / 0.14 sq. ft.	1,786

Table 7-5: Pipe Velocity in a 250 ACFM Application

As you can see, and probably would expect, the velocity of air decreases rapidly with increasing pipe diameter. In this case, it's not until 3″ diameter pipe is reached that the velocities drop below the 6,000 feet per minute threshold. Just by looking at the table, you can answer the questions of whether or not 2″ diameter pipe is too small and if can it be used in this application. The answer is yes to the first part of the question and maybe to the second part. The reason "maybe" is the answer to the second part of the question is that even though the velocities (and pressure drop) may be too high, there is still vacuum being drawn through the pipe and the process can possibly be run. The real question is will the users be getting their money's worth out of the vacuum pump?

The other question that can be asked here is this: can the pipe diameter be too big for an application? The answer is not typically. Larger pipe diameters in most rough vacuum applications will not affect the operation of the vacuum pump or hurt process delivery. In some cases though, such as cyclic applications where every additional cubic foot of volume adds to the pumpdown time, larger-than-required supply piping can be a problem.

These are the exceptions rather than the rule however. In many cases, there are benefits attached to having slightly larger diameter vacuum pipe for the simple reasons of planning for plant expansions and lower pressure drop. As was already mentioned, plant expansions or production equipment additions can create havoc with the vacuum supply piping. These problems can be avoided if the initial piping system is designed and installed with the intention of adding on new equipment.

The next most important question to ask when designing a vacuum supply system is: which piping configuration is best? It is very important to have short, straight piping configurations in vacuum supply systems, especially if piping diameters are marginally sized. While it is impractical in most applications to run a piping system in a straight line from the process to the pump or pump room, there are many applications where the vacuum supply piping appears to have no discernible form. Many of these systems can be simplified and cleaned up with minimal effort. The economic results for doing this can be tremendous. As discussed earlier, pressure drop in vacuum supply system creates the need for additional capacity from supply pumps. Maintaining low system pressure drop is critical for good vacuum delivery to production machinery as well as to prevent operation of unneeded vacuum pumps.

The reason vacuum supply piping should be as straight as possible and as short as possible is that every 45° or 90° turn adds what is known as **"equivalent length"** to the piping system. Equivalent length is simply the amount of straight pipe that would have to be added to a piping system to generate the same amount of pressure drop as the corresponding change in direction would.

For example, if a vacuum supply system has 5" diameter pipe, then every 90° turn adds about 14 feet of equivalent length to the calculation for pressure drop. If a particular system has 150 feet of 5" pipe and twelve 90° turns, then the total calculated pressure drop is for 150 + (12 × 14) = 318 feet of pipe. This provides ample motivation to keep the system piping as straight as possible. With each added turn or elbow, the pressure drop increases in response. One thing to keep in mind, however, is that the pressure drop added from 90° turns can be minimized by using large radius elbows instead of straight 90° turns.

Figure 7-1 illustrates the difference between a large radius elbow and a straight 90° turn.

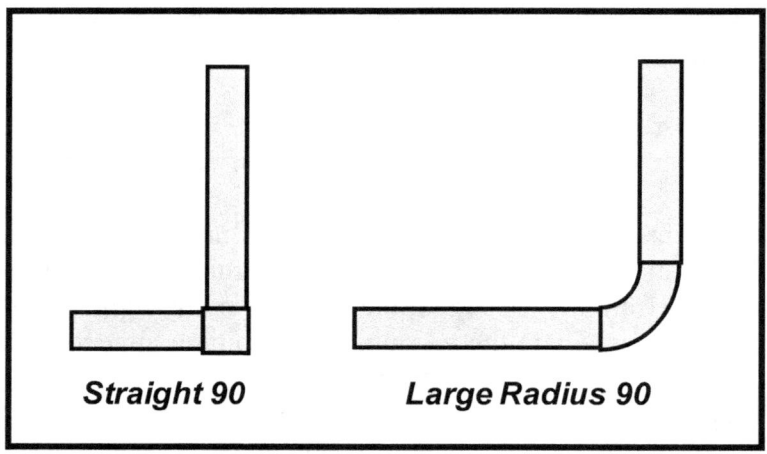

Figure 7-1: Straight 90 vs. Large Radius 90

Using large radius elbows can significantly reduce the amount of added pressure drop from abrupt changes in direction. When designed and installed properly, a good vacuum distribution system will be well worth the effort. Good designs will pay for themselves over many years of operation through savings in energy costs and reductions in maintenance loads.

One very common central vacuum system piping configuration is the loop. A loop encircles the production floor and has one trunk line connection to the vacuum pumping system. See Figure 7-2. One frequently asked question about vacuum piping configurations is: should the piping be looped in a circle or is it better to run a straight trunk line down the middle of the production floor? The answer is that they are both viable systems if designed and installed properly.

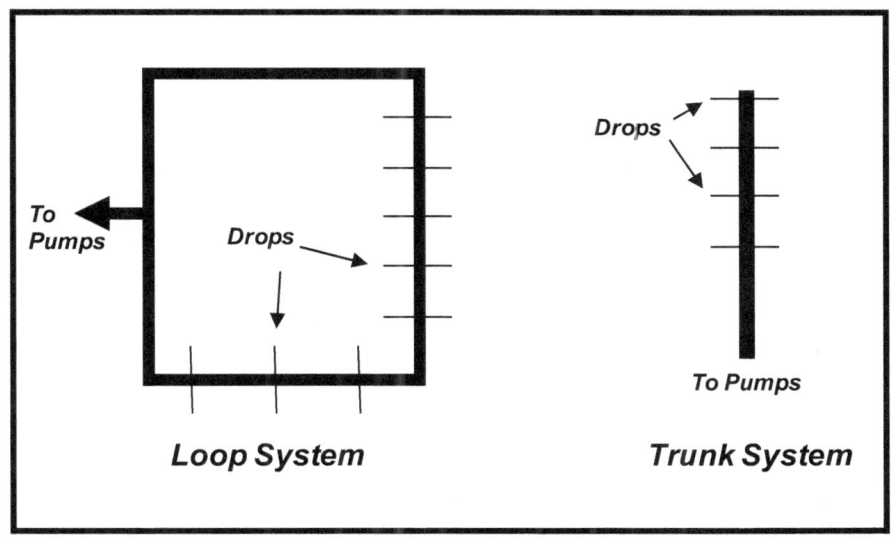

Figure 7-2: Loop System vs. Trunk System

Loop systems typically have greater physical volume than comparable trunk line systems. When properly sized pipe diameter is used in a loop system, there can be a significant buffer available for system upsets. When additional physical volume is present in a vacuum system, any extra air mass that enters the system from leaks or process upsets will have a smaller effect on the vacuum level than there would be on a comparable system with smaller physical volume. This concept can be useful if there are frequent process interruptions. A good example would be instances where production machinery leaks large volumes of air into the vacuum system during work stoppages.

In some cases, it is difficult to install a trunk line system due to space restrictions resulting from the larger diameter pipe. Loop systems can provide the same delivery and pressure drop performance as trunk line systems with smaller diameter pipe. The reason is because the total flow is split in half. One half of the process flow is being pulled through each side of the loop system. In most installations, it is necessary to look closely at the way process machinery is placed on the production floor before decisions can be made regarding the most convenient supply system. In many cases, large loop systems will have smaller diameter "crossover" pipes that connect the two halves of the loop in various locations. These crossovers help reduce pressure drop in localized sections of the system.

Use of the correct materials of construction is very important when designing and building a vacuum supply system. There are three issues associated with materials of construction that must be addressed. The first issue is process compatibility. Each potential gas, vapor, liquid or particulate must be reviewed to ensure that it is compatible with the materials of construction used in the system piping and in any in-line accessories.

For example, PVC is a common material of construction used in many rough vacuum piping systems. It is chemically resistant to many liquids and vapors and holds up well under reasonable vacuum levels. However, there are common process chemicals such as acetone and some forms of alcohol that will attack and/or corrode PVC. In these cases, PVC is a poor choice for the system piping. Carbon steel is also widely used for vacuum system piping but it, too, can be the wrong choice in some applications due to the risk of corrosive or abrasive attack. The bottom line is that with good information regarding the process, compatible materials of construction can be chosen.

The second concern about materials of construction is adequate sealing characteristics. Will the material be able to hold up, long term, in vacuum service without developing excessive leaks? Various types of flexible hoses are used to connect vacuum pumps to their respective processes. There is a constant flex on piping that results from being under vacuum and cycling back to atmospheric pressure. Repeated flexing can result in the development of cracks in hose walls. When these cracks become large enough, air will leak into the system and be the cause of poor vacuum delivery at the process.

Other types of hoses can completely collapse, cutting off or severely reducing vacuum supply. Some types of rigid plastic pipe can implode when exposed to the low pressures associated with vacuum. In these cases, even though the mode of failure is implosion, what actually happens is that the pieces fly inward toward one another, hit and bounce back out in an exploding fashion. This is a dangerous situation and can easily be avoided by consulting specifications on the particular materials being proposed. Other materials can be fine for many years, then develop hairline fractures. This can result in the same type of failure. Be sure to choose materials that have good threaded, glued, welded or hose type connections. Many times the connection points between two adjacent lengths of pipe or tube become the weak point in the system.

The third concern about materials of construction in vacuum supply piping is pressure drop. The absolute best material for pressure drop is glass because of its' smooth interior surface. Smoother surfaces result in lower air friction along the walls of a pipe. Of course, most rough vacuum systems cannot use glass piping, but in general, the point is that the rougher the finish, the greater the pressure drop. For obvious reasons, the problems gets much worse as pipe length increases. PVC and other plastics have very smooth interior surface finishes and do very well in vacuum service.

Probably the two most common materials of construction used in industrial vacuum piping systems are carbon steel and PVC. They are both resistant to many different gases, liquids and vapors. They have great sealing capabilities and exhibit low or moderate pressure drop in most systems. PVC is used commonly because it is very light, easy to install and in most areas, very reasonably priced. With PVC, it is typically not the cost of the materials that is most significant. It is the cost of the labor to install it.

Accessories

Now is a good time to look at the various types of accessories used in rough vacuum systems. Accessories such as vacuum receivers, check valves, isolation valves and vacuum gauges are all commonly used in vacuum systems to assist in the delivery of vacuum to meet production needs. These items provide assistance by either protecting the vacuum pump or providing information on the condition of the vacuum system. Filtration, both inlet and discharge, is covered in detail in Chapters 5 and 6.

The discussion on vacuum accessories begins with the use of vacuum receivers. There are many common misconceptions regarding the use of vacuum receivers and their viability in vacuum systems. Receivers are normally used for liquid separation, as short term high capacity boosters and as system buffers. As a rule of thumb, in most systems, it is generally not recommended to use receivers for vacuum storage, but there are some systems where storage receivers are useful and necessary. The reason most systems do not benefit from vacuum storage is easy to understand when you think about vacuum storage in terms of pressure. A majority of continuous operation vacuum systems operate within a vacuum range of 2″ Hg to 4″ Hg above or below a target pressure. When systems operate outside of that range, process and/or production machinery will slow down or stop due to inadequate vacuum supply. When these conditions are equated to PSI, 2″ Hg to 4″ Hg equals 1 - 2 PSI. This is very little pressure differential in terms of a range in which to store vacuum (or even compressed air for that matter).

As an illustration of how inefficient it is to store vacuum in a production environment, an example system will be used. This example system has a 300 ACFM (20 HP) vacuum pump operating on a production system that requires a vacuum level of 18″ HgV. Table 7-6 below illustrates the amount of time various sized receivers can be used to pick up the production load (within a 4″ Hg range). Note that the receiver will have to be pulled down to 22″ HgV because of the 4″ Hg differential pressure requirement.

Demand Capacity	Rec. Size (Gal.)	Time to Depletion
120 SCFM	100	0.9 seconds
120	200	1.8 seconds
120	400	3.6 seconds
120	1000	8.9 seconds
120	5000	44.5 seconds
120	10000	1.5 minutes
120	30000	4.4 minutes

Table 7-6: Depletion Time in Seconds for a Receiver

Notice that a 10,000 gallon tank provides only 1.5 minutes worth of storage time for this process. This is a very large storage receiver for such a small vacuum system. Also note the very small amount of storage capacity (3.6 seconds) for a reasonably sized storage tank of 400 gallons. For the amount of money spent on a storage tank, the payback for most vacuum systems is just not available.

Having illustrated that, there are instances where stored capacity can be used very effectively. The first are applications that require very high capacity for very short periods of time. This is quite common in plastic thermoforming applications. Large receivers are used in these systems as "vacuum pumps" with high capacity. The receivers are pumped down to the deepest vacuum attainable in the amount of time allowed between cycles. When the forming table is ready with the hot, flat sheet of plastic placed over the mold, the isolation valve is opened to the storage tank. The tank is at full vacuum and the table or mold is at atmospheric pressure. When the valve is opened, there is almost an instantaneous equalization of pressures.

Vacuum pulls the plastic sheets into the shape of the mold and holds them there until they cool and solidify into the desired shape. The isolation valve is then closed and the receiver is pulled down to full vacuum again. When the cycle timing is correct, this is a very effective method to avoid having an enormous vacuum pump to pull the hot plastic sheet into the mold. The following discussion provides the formula and an example on how to size a receiver for these types of applications.

RECEIVER SIZING

This explanation of how to size a vacuum receiver for the rapid pumpdown of a vacuum chamber at atmospheric pressure to a predetermined level of vacuum is a general description only. Please consult with a qualified engineer when designing systems for real life applications. The following base formula is used:

$$(P_1 \times V_1) + (P_2 \times V_2) = (P_3 \times V_3)$$

(also written as: $P_1V_1 + P_2V_2 = P_3V_3$)

Where:

P_1 = the initial chamber pressure

P_2 = projected receiver pressure (lowest)

P_3 = intended average pressure in the system after the isolation valve

is opened between a process chamber and the storage receiver

V_1 = the process chamber volume in cubic feet

V_2 = receiver volume in cubic feet

V_3 = process chamber volume in cubic feet + receiver volume in

cubic feet

In general, the goal is to isolate V_2 in the formula so the correct size receiver can be determined for the application. The math is broken down step by step so it is easy to follow along and determine V_2.

Starting with:

Step 1: $(P_1 \times V_1) + (P_2 \times V_2) = (P_3 \times V_{3)}$

Step 2: $(P_1 \times V_1) + (P_2 \times V_2) = P_3 \times (V_1 + V_2)$ [since $V_3 = V_1 + V_2$]

Step 3: $(P_1 \times V_1) + (P_2 \times V_2) = (P_3 \times V_1) + (P_3 \times V_2)$

Step 4: $(P_1 \times V_1) - (P_3 \times V_1) = (P_3 \times V_2) - (P_2 \times V_2)$

Step 5: $(P_1 \times V_1) - (P_3 \times V_1) = V_2 \times (P_3\text{-}P_2)$

The following equation results and is utilized to solve the problem:

Step 6: $V_2 = V_1 \times (P_1\text{-}P_3) + (P_3\text{-}P_2)$

In essence, this is a formula where all the known variables can be inserted on one side of the equation so that receiver volume can be determined. To illustrate how this works, please work through the example problem below.

EXAMPLE:
An end user has a 15 cubic foot total mold volume for a plastics forming process that forms refrigerator doors. They need to pull the mold down to 25" HgV as quickly as possible to run this process most effectively. There is a central vacuum system operating at 28.5" HgV and it is this central system that will be connected to the production process. What size receiver should be used in this application to attain the goal of rapid pumpdown?

Once again, the formula is:

$V_2 = V_1 \times (P_1\text{-}P_3) + (P_3\text{-}P_2)$

In this case, P_1 is atmospheric pressure, or 760 torr, which is the starting pressure in the mold. V_1 is the mold volume which is 15 cubic feet. P_2 is the average pressure in the central vacuum system and will also be the pressure in V_2, the storage receiver. This pressure is 28.5" HgV or 36 torr (remember the conversion to absolute pressure). P_3 is the desired end pressure after the isolation valve is opened and the pressures equalize between the receiver and the mold.

When the chamber, which is at atmospheric pressure, is opened to the receiver and central vacuum system, there will be a rapid equalization of pressures and the target pressure of P_3 will be obtained. The target pressure in this case is the desired end pressure at the mold which is 25" HgV or 125 torr. Note that absolute pressure terms are used in all vacuum formulas! When the values are placed in the formula, the resulting formula and answer is:

$V_2 = 15 \times (760\text{-}125) + (125\text{-}36)$

$V_2 = 107$ Cubic Feet Volume (800 Gallons)

If you would like to check your work, put the receiver volume back into the original formula and see if the end pressure is the desired end pressure:

$(P_1 \times V_1) + (P_2 \times V_2) = P_3 \times V_3$

$(760 \times 15) + (36 \times 107) = P_3 \times 122$

$15,252 = P_3 \times 122$

$125 = P_3$

The answer turns out to be correct. Keep in mind that there are many factors that can affect the actual time it takes for the pressures in the two volumes to equalize. Air travels at fairly high velocities under vacuum and factors like piping, valves, elbows, leaks and restrictions will alter the actual results from the theoretical results - so be careful!

The second common application for receivers is to smooth out pressure fluctuations in production vacuum systems. Certain systems operate at steady vacuum levels until process events occur that upset the system. If the vacuum pump in place is not adequate to handle the additional air flow from the process upset, the vacuum level in the system can fluctuate dramatically unless an adequate buffer is available. Additional physical storage volume will smooth out fluctuations in the vacuum level which results in a smoother process profile. In any case, when a receiver is put into a system, the system vacuum pump must be sized so that there is extra capacity available to not only handle process demand but also to pump out the additional storage volume in the system. The extra capacity is required to pull the system back down to an acceptable vacuum level before the next process upset occurs. If the vacuum pump is not large enough, the system will not be able to recover to the original vacuum level and vacuum supply to the process could be hindered.

The third most common method where receivers are used effectively in vacuum systems is the removal of liquids from the incoming gas stream prior to the inlet of a vacuum pump. Serious damage can occur to most types of vacuum pumps if liquids are ingested in high quantities. This is due to the fact that liquids are not compressible. When high volumes of liquid enter a vacuum pump, there may be an insufficient amount of air available to buffer the compression chamber as the pump goes through its compression cycle. The damage occurs as the vacuum pump tries to compress the volume of liquid in the pumping chamber. To avoid this potentially harmful circumstance, a separator or some other type of receiver can be placed prior to the inlet of the vacuum pump. Refer to Figure 7-3 below for some of the common types of separators used in vacuum systems.

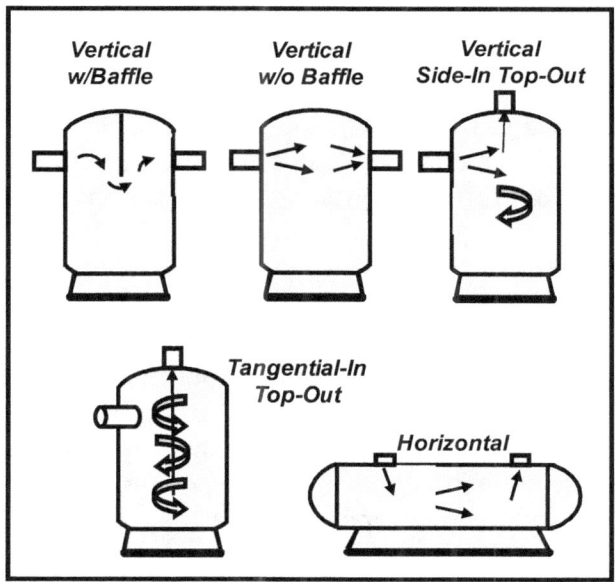

Figure 7-3: Typical Receiver Configurations

There are many other types of separators available which will assist in the removal of liquids from a vacuum air stream. These listed above, however, are the most common. It is important to keep pressure drop to an absolute minimum when considering the installation of a receiver for inlet liquid separation. For example, of the receivers listed above, the horizontal receiver would be a bad choice for an application where there is a heavy liquid load. High liquid level can block the inlet or discharge port on the receiver. As it begins to fill up with liquids, the pressure drop across the receiver will increase. The reason for this is the available area for air flow has been reduced. In this particular case, rising liquid level creates additional air friction and higher velocities.

Other types of receivers can exhibit high pressure drop as well. If the diameter of the inlet and exhaust porting on any of these designs is not large enough, high pressure drop will be the result. Many standard compressed air receivers are not good choices for vacuum service. Any sudden enlargement or sudden restriction in vacuum system supply piping will cause an increase in pressure drop. Typically, receivers have both a sudden enlargement and a sudden restriction.

One contaminant that standard receivers will typically not remove from an incoming gas stream is vapor. Water vapors, solvent vapors and other similar materials are usually carried right along with the gas stream and are difficult to remove with an in-line receiver. The reason is that these vapors are actually part of the gas stream and are not considered to be carried along in the gas stream like liquid mist and solid particles. The only effective way to remove vapors with a receiver is to cool the outside (hence the inside) surface with chilled water or some other refrigerant so that vapor molecules condense on the cool interior surfaces. After condensation, the resulting liquid can be drained from the system. Depending upon the amount and type of vapor and the inlet gas stream temperature, a receiver may or may not act as an effective vapor trap.

A better way to separate vapors from the incoming gas stream is with an inlet condenser similar to a shell and tube design heat exchanger. Other than vapors, receivers will also have a difficult time with light, dry particulate. The turbulent air inside the receiver will pick up the light particulates, spin them around and eventually carry them out of the receiver and into the vacuum pump. Heavier or wet particulates are much better separated with a receiver than the light dry type. One last thing to note about receivers. When purchasing or designing receivers for vacuum applications it is recommended that they have some type of clean-out port so that the inside of the receiver can be occasionally cleaned and/or inspected. There is a design code for receivers and, in general, it reads that they should be rated for full vacuum. Check with the receiver supplier to ensure that it meets any required local codes.

Check valves are also a common accessory used in vacuum systems. Check valves are used on the inlet of vacuum pumps for three primary reasons. The first is to ensure that when a liquid sealed vacuum pump is shut down, there is no possibility for the pump oil or water to be drawn out of the vacuum pump and into the vacuum system (or other vacuum pump on the same system). When a liquid sealed vacuum pump is shut down, there can be a direct path through the pump that leads from the discharge port, where atmospheric pressure is present, through the pump itself and into the vacuum system. After the vacuum pump is shut down, atmospheric pressure at the discharge port, coupled with system vacuum on the inlet port, can create a condition where the lubricating or seal fluid is pushed from the discharge side of the pump into the vacuum system. Liquids from the vacuum pump can be drawn into the production vacuum system, or if there are other vacuum pumps in operation on the same system, fluids can be drawn from one vacuum pump and into another.

This situation can get especially tricky when water sealed pumps are used in conjunction with oil sealed pumps. Installation of an inlet check valve will prevent this situation from occurring, because the volume of space between the vacuum pump inlet and check valve is usually not very large and any air or oil leakage from the discharge side of the pump quickly brings this small volume back up to atmospheric pressure. Therefore, there is no "pull" on the inlet side. In addition, some users will place an automatic vent valve on the pump side of the check valve that opens when the vacuum pump shuts down. This quickly raises the pressure back to atmospheric and prevents any back migration of liquid.

Another reason for using a check valve on a vacuum pump is to prevent the back flow of air into the vacuum system through a vacuum pump that is shutdown. This situation occurs for the same reasons as those listed above. Some vacuum pumps actually have a device built-in to the pump itself to prevent this from happening. These devices, sometimes called **anti-suckback valves,** are designed and built into the vacuum pump so that external check valves are not necessary. Some of these valves are spring loaded, while others use system vacuum to hold the device closed.

The last common reason for using an inlet check valve is to eliminate the system inlet isolation valve. If the check valve is of sufficient quality, it will be able to hold atmospheric pressure on one side and full vacuum on the other without leaking. Some users install only check valves in their vacuum systems and forego the installation of isolation valves. In systems where there is little danger of either contaminating a process or of losing vacuum, this is an acceptable practice. However, it is recommended that if production will suffer because of vacuum loss or if liquid contamination from the pump has the capacity to damage production equipment or other vacuum pumps, separate isolation valves should be installed. Typically, check valves that are stuck in the open position cannot be overridden and closed without a significant effort. If the check valve fails for any reason, a separate isolation valve is a good form of insurance. As a side note on check valves, the best designs for vacuum are typically elastomer sealed and spring loaded. Metal to metal seals are not as forgiving when particulate is present and if improperly sized, can "chatter" under certain flow conditions. Metal sealed swing type check valves tend to chatter in many vacuum applications.

There are many different types of isolation valves available on the market and most of these will work in rough vacuum applications. There are a couple of things to keep in mind when purchasing and installing an isolation valve on a vacuum system. First, the valve should be rated for vacuum service and have a leak rate specification for vacuum. The leak rate spec should be a very small fraction of the total capacity of the vacuum pumping system. Many valves that are sufficient for low pressure compressed air may not be suitable for vacuum because the method of sealing is inside out versus outside in. The wrong type of valve can leak and cause problems when the pump is shut down or off line.

The second item to note is that the more the valve port resembles the diameter of the supply piping the better. Inside diameters of standard port ball valves typically do not match the inside diameter of the pipe connection and can cause a significant amount of pressure drop in a system. The best types of valves to use in vacuum applications are full port ball valves, butterfly valves or full port gate valves. As with any accessory installed in a vacuum application, be sure that the materials of construction are compatible with the process gas stream. A process gas stream that attacks the elastomer seals in an isolation valve can cause many unseen and frustrating problems with leaks and contamination.

Vacuum gauges are the next accessory on the list. The most important question regarding most vacuum systems is the operating level of vacuum. Having said that, it is typically the least known variable in many vacuum systems. Since there is so much confusion with users about the level of vacuum in their systems and that much of this confusion revolves around the accurate reading of a vacuum gauge, please refer users to the pictorial guide below on the five most common vacuum gauges and the proper way to read them. All of the pictured gauges have atmospheric pressure and full vacuum labeled for ease of understanding. The five scales used are the most common found in rough vacuum applications: PSIA, Inches of HgV, Inches of HgA, Torr and Mbar. The most important (yet most confusing) question you can ask vacuum users is what scale are they are referring to when speaking with them about their vacuum application.

Many users will not know or will make an assumption about which scale they are using. It is important to have accurate information regarding the vacuum level because it is the foundation of your understanding of the system for pump sizing, pressure drop issues or when application problems are being discussed. Feel free to copy this page and use it as necessary.

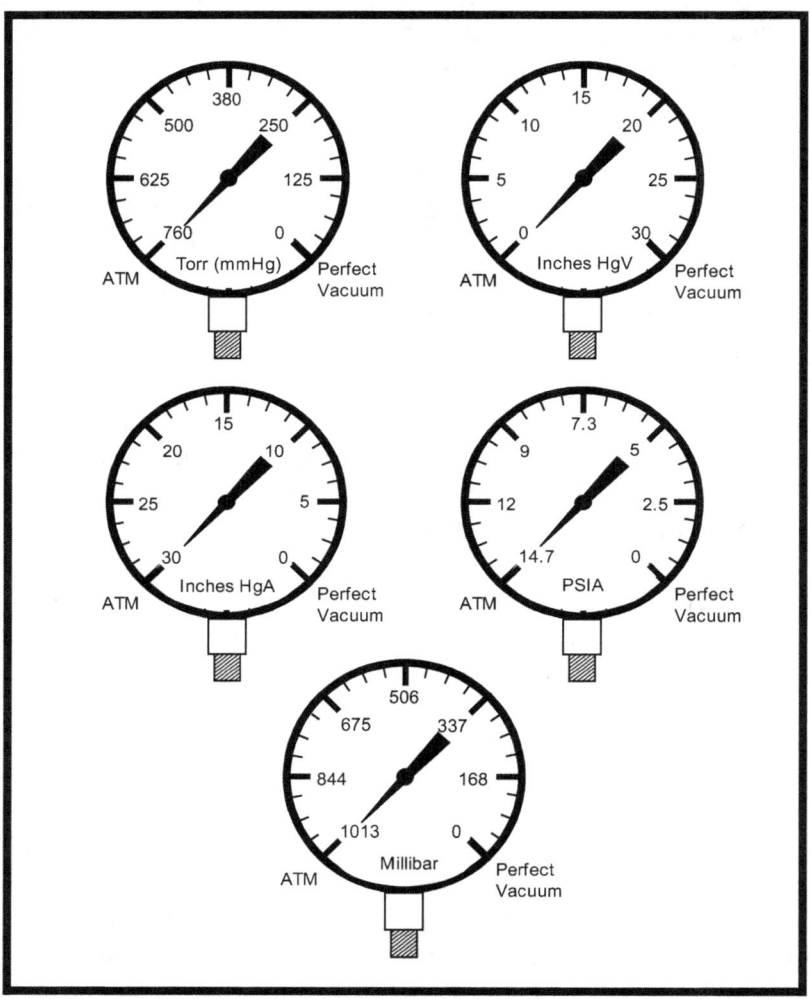

Figure 7-4: Typical Vacuum Gauges

The second most important aspect of installing and/or reading a vacuum gauge in a given vacuum application is placement. As confirmation to the discussions on pressure drop, a vacuum gauge mounted at or near the vacuum pump will typically have a different reading than the gauge mounted at the process or point of use. Of course, the reason for different readings is because of the pressure drop that exists in the distribution system. When working with vacuum equipment users over the phone or in person, it is critically important to establish at what point the vacuum is being read so that correct decisions can be made about the system. Many users will not know the importance of gauge placement and proper reading of the vacuum gauge. It is very important to help them understand why you are asking this question.

The most common vacuum gauge used in standard rough vacuum applications is the Bourdon Tube vacuum gauge. This vacuum gauge is inexpensive, reasonably accurate for the applications where it is used and durable enough to last for years in many different installations. The Bourdon Tube is actually a coiled hollow tube with one end open to the vacuum system and closed on the other end. As vacuum is drawn on the system, the coiled tube coils further in response to the changing pressure. There is a mechanical connection from the tube to a lever that moves a needle around the dial face. As deeper vacuum is drawn, the interconnecting lever moves the pointer needle in response, and the vacuum level is known. Not all vacuum gauges are alike however.

Just like buying any other instrument or piece of process equipment, you get what you pay for. As you learned earlier, there is no way to attain perfect vacuum here on earth. There is a saying in the vacuum industry that the less money you spend on a vacuum gauge, the greater the likelihood you will attain perfect vacuum! The point is, spend money on the level of quality and the level of resolution needed for a particular application.

One of the problems with this type of gauge is operation at elevations above mean sea level. The higher the elevation above sea level, the lower the ambient atmospheric pressure and subsequently the smaller the deflection of the gauge needle. This is why Bourdon Tube vacuum gauges in applications at high elevations will not be able to display 29″ HgV or 29.5″ HgV for example. It is because they are built and calibrated for atmospheric pressure at sea level and cannot deflect in response to vacuum at elevation as much as they deflect at sea level. Another problem is vibration. These gauges are sensitive to the vibrations associated with some types of vacuum pumps. In those installations, liquid filled gauges can be used to dampen the bouncing of the indicator needle. See Figure 7-5 below for a diagrammatic view of the workings of a Bourdon Tube vacuum gauge.

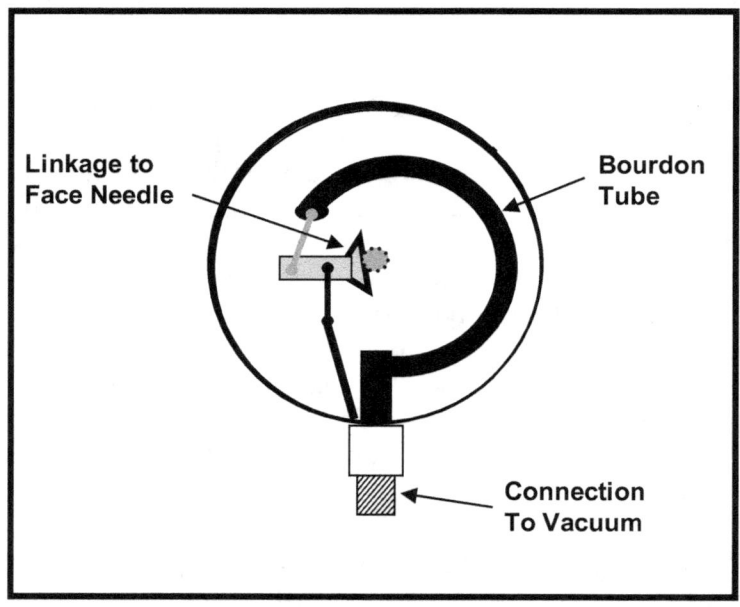

Figure 7-5: Internal View of a Bourdon Tube Vacuum Gauge

Some other types of vacuum gauges you will commonly see in rough vacuum applications are listed below with brief descriptions on how they operate.

Pirani and Thermocouple Gauges: These vacuum gauges are electronic type gauges that either measure the resistance of electricity across an element or they measure the heat generated by the element. Both operate on the electronic principle of the Wheatstone Bridge. These gauges are typically absolute pressure gauges in that they do not use atmospheric pressure as a reference. While these gauges are used extensively in industrial applications, they are mainly used in the lower end of rough vacuum and well into the medium vacuum range.

U-Tube Manometers: A very simple type of vacuum gauge that uses a "U" shaped tube filled with a liquid. The liquid can be virtually any material but water, however, certain types of oils are used most commonly. One end of the tube is connected to the vacuum system, while the other end is open to atmospheric pressure. Graduations on each side of the tube are measured and added together to obtain the vacuum level in the system. This gauge is not considered an absolute pressure gauge.

Diaphragm Gauges: Diaphragm vacuum gauges work in a similar manner to Bourdon tube gauges, except that instead of a bourdon tube, a flexible diaphragm responds to the changing vacuum in a system. Diaphragm gauges are used in many process or chemical applications because the materials of construction exposed to vacuum can be easily controlled and process compatibility can be maintained. There is typically a higher level of accuracy associated with this type of gauge. The diaphragm gauge is considered an absolute pressure gauge because it uses a reference vacuum in a sealed chamber within the gauge instead of using atmospheric pressure as a reference pressure.

Capacitance Manometers: This gauge is very similar to the diaphragm vacuum gauge in that it uses a vacuum in a sealed chamber as a reference pressure instead of atmospheric pressure. The difference is that the capacitance manometer does not use a mechanical lever system to respond to and read pressure levels. Instead, it measures vacuum level by sensing the change in electronic capacitance between two points, one point on the flexible diaphragm and one point that is stationary in the gauge body itself. These gauges are absolute pressure gauges and are typically very accurate.

Mercury McLeod Gauge: While very similar in operation to the U-tube manometer, the McLeod gauge can measure vacuum with greater accuracy. This gauge uses mercury instead of oil or water and can be read very accurately well into the medium vacuum range.

One of the variables in many vacuum systems is the amount of water vapor present. The McLeod gauge reads mainly permanent gases because many of the vapors present will condense on the surface of the mercury.

In most standard rough vacuum applications the Bourdon Tube is the gauge of choice because of its overall versatility and low cost. In any vacuum system, there should be ports available in which to install vacuum gauges at various locations within the distribution piping and at the supply pumps. It is a good idea to have small isolation valves mounted with the vacuum gauges so that replacement can be done without process upsets. As always, make sure the materials of construction are compatible with the process gas stream.

In general, many of the pressure drop issues associated with rough vacuum systems can be solved by looking closely at the distribution system. High profile items like supply piping, vacuum pump accessories and receivers are many times put together without a cohesive, systematic approach to design. Any one of these items could be the major contributing factor in poor vacuum delivery, but it is more likely a combination of many small problems that add up to one large issue. To keep out of system delivery trouble, it is necessary to study the design and placement of each component so that when they are placed in the system, they work together effectively.

Chapter 8

Water and Water Vapor

Water is present in one form or another in every vacuum system here on earth. Even the highest-tech space simulation chambers used in government research facilities have water vapor present – and no one can completely get rid of it! The following discussions regarding water and water vapor in vacuum systems were placed at this point in the book because a thorough understanding of vacuum systems and a familiarity with the terminology associated with vacuum is required. Not only is water the curse of high vacuum systems, but it is also a problem in many rough vacuum systems and can destroy vacuum supply pumps slowly over time or instantaneously in one single event. Therefore, an understanding of how to remove the various forms of water from an incoming gas stream is a necessary skill. This chapter is divided into two parts. The first section covers liquid water while the second section examines water vapor.

LIQUID WATER

Liquid water is present in a vast number of rough vacuum applications. Drying, plastics processing, food processing, pulp and paper and a host of other applications all have liquid water present in varying amounts. When liquid water enters a vacuum system, it will typically do one of three things: stay at the point where it entered the system; travel through the vacuum piping and either enter the vacuum pump or remain in the system piping/receiver; or it will flash into a vapor and become part of the gas stream. If water gets into a vacuum system and collects at a certain location, it can eventually evaporate and become part of the vacuum gas stream. The biggest problems occur when liquid water enters the system and is drawn back to the vacuum pump.

When liquid water enters a vacuum pump there can be a range of reactions anywhere from nothing happening at all to a complete failure of the system. The reaction to water ingestion is dependent upon the type of vacuum pump. Water-sealed liquid ring vacuum pumps are probably the most forgiving technology overall because water is used as the sealant between the impellers and housing. When reasonable amounts of water are ingested into a water-sealed liquid ring vacuum pump, the additional water is typically swept away along with the normal seal water flow. Oil-sealed vacuum pumps, however, are typically not that forgiving.

In general, when large quantities of water are taken into a vacuum pump, some of the possible results are:

Minor Damage: Several minor occurrences can damage the pump. For example, couplings can break, contaminated oil will need changing, air/oil separators and filters can be contaminated and require replacement and internal pump components can rust or corrode.

Moderate Damage: Electrical drive motors fail due to excessive loading, vane breakage in the pumping modules of rotary vane vacuum pumps, bent or damaged exhaust valves, deformation of internal pump components from hydraulic pressure, bearing failure over time due to contaminated lubricant, sludge build up from contaminated water or a reaction of the pump sealing/lubricating fluid with water.

Severe Damage: Complete destruction of the pumping module due to the hydraulic pressures associated with the attempt to compress a liquid.

With these types of potential pumping system problems, it is important to remove liquids before they get into a vacuum pump. Liquids entrained in the incoming air must first be removed from the air stream, trapped in a reservoir and then drained from the system without disrupting production vacuum supply. There are some general guidelines to follow when working with pumps that could potentially ingest liquid water. The following list provides brief descriptions of each guideline.

Receivers: It is critical to have a liquid receiver prior to the inlet of a vacuum pump so that liquid water can be removed from the air stream and stored prior to removal. The type of separator or receiver used will depend upon economics and the amount of water that could potentially enter the system. In applications where there are small amounts of liquids present, standard vertical compressed air receivers are installed at the inlet of the vacuum pump. The only issue with these types of receivers is that there is typically no change in the direction of air flow. The air flow comes straight in one side and goes straight out the other. When there are changes in the direction of air flow, the larger water droplets tend to separate from the air stream. A baffle plate placed between the inlet and discharge ports on vertical receivers works very well. Separators with tangential side inlets and top outlet ports work equally well.

Piping Headers: Any long stretches of vacuum piping should slope downhill toward a receiver or separator. This way, water will not collect in system low spots. If there is no receiver or separator in the system, there should be drop legs with drain valves placed at low points in the system piping. These will assist in the removal of liquids that have been collected. The buildup of water or other liquids in vacuum piping can cause an increase in pressure drop across that section of pipe. It is recommended that these locations be placed on regular preventative maintenance programs.

Materials of Construction: Make sure water will not rust or corrode system piping over time. Plastic, galvanized metal or other metals that are resistant to corrosive attack will work well in these applications. If local codes allow, PVC is an excellent choice for liquid water applications.

Start-Up and Shut-Down Procedures: With many types of oil sealed vacuum pumps, it is a good idea to start the vacuum pump twenty minutes or so prior to opening the inlet of the pump to process air flow. The reason for doing this is that it may take a typical oil sealed vacuum pump about that long to build up to normal operating discharge temperature. If process air flow is run through the vacuum pump prior to the discharge temperature being hot enough, water that enters the vacuum pump can become entrained in the lubricant. If the discharge temperature is high enough, water has a chance to exit the vacuum pump as a vapor and then be ejected from the system. The same is true about shut down. Before the vacuum pump is shut down at the end of the day or shift, it should be run for several minutes with a non-process dry air flow. This is so any entrained water has a chance to be ejected from the pump.

Once the liquid water is trapped in the separator or receiver, it must then be drained from the system. If there is only a small amount of water present and the plant is not running 24 hours/7 days a week, the receiver can be drained at the end of the shift, day or week by simply opening a drain valve when the vacuum system is off. If the amount of water is sufficiently large and/or the system is running full time, the stored water must be drained during live operation. To drain water from a live system, a second storage tank may be necessary. It is rather difficult, however, to drain water out of a live vacuum system. If a normal drain valve on a vacuum receiver is opened while the system is under vacuum, there is a high probability that there will be more air entering the system than water leaving the system. To drain water out of a live vacuum system, one of three general conditions must be met:

1. A separate storage receiver that can be isolated from the main separator or receiver must be in place. See Figure 8-1. Liquid water gravity drains into the secondary tank during normal operation. When the secondary tank fills to the point where removal of the liquid is necessary, it is isolated from the main system. An air inlet valve is opened to equalize the pressure back to atmospheric and then a drain valve is opened to release the water.

After the water is drained from the secondary tank, the valve sequence is reversed with special caution given to the main isolation valve. The air in the secondary tank, now at atmospheric pressure, must be slowly admitted to the primary receiver. Otherwise, water that has accumulated in the bottom of the primary receiver can be swept out of the receiver and into the inlet of the vacuum pump. Also, too much air entering the system will lower the system vacuum level.

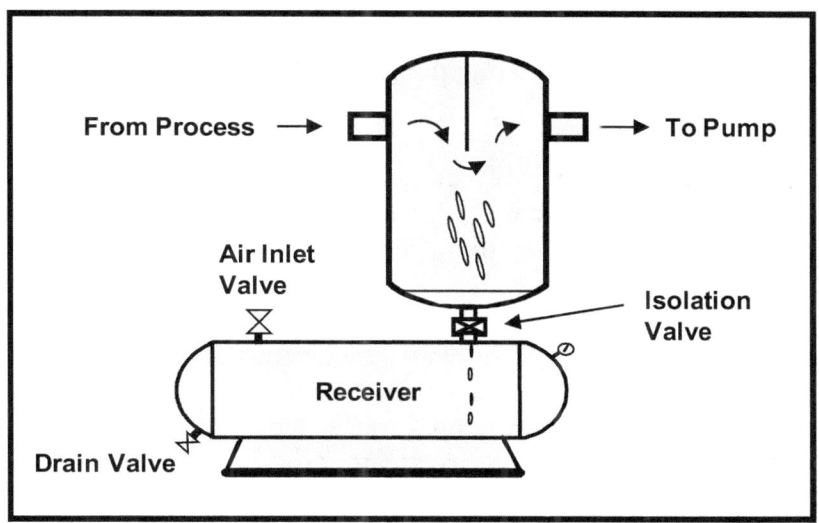

From Process → | | To Pump

Air Inlet Valve

Isolation Valve

Receiver

Drain Valve

Figure 8-1: Secondary Drain Tank for Water

2. A drain leg must be in place so that the weight of water in the drain leg is greater than the force of vacuum pulling it up. At the very bottom of the drain leg, an isolation valve or float valve can be added so that water can be periodically removed from the system. Under full vacuum conditions, the drain leg would have to be approximately 34 feet tall. Similar systems, called barometric legs, that drain vacuum systems automatically, can also be designed and installed.

3. A water pump with a low, net positive suction head (NPSH) must be installed along with a float switch that automatically pumps the system out when the receiver accumulates a prescribed amount of water. The liquid pump overcomes the pull of vacuum and removes the water from the storage system as it accumulates. This type of system negates the need for a secondary receiver because the liquid pump removes water directly from the primary receiver or separator.

Typical liquid receivers and separators do a good job of removing large water droplets and particles, but for finer mists other separation techniques may have to be employed. The efficiency required for water removal depends on the tolerance of the vacuum pumping system for water contamination. If the vacuum pump can take virtually no water contamination, then higher efficiency systems must be employed.

The mist removal efficiency of standard receivers can be improved by adding more baffle stages, demister pads or both. Velocities also play an important role in the effectiveness of a separation system. Since fine liquid mist is carried along with the gas stream, it becomes increasingly difficult to remove the smaller diameter mist particles. Additional baffle stages and demister pads increase the probability that mist particles will come in contact with a solid surface. When they hit a solid surface, they can either bounce off and remain entrained in the gas stream, or they can stay in place and coalesce into larger particles. Once liquid particles are large enough, they have the capacity to be gravity drained from the system.

One important point about liquids in vacuum systems. Once liquids are separated from the incoming gas stream it is important to be able to get those liquids out of the system as soon as possible. This is especially true in high volume systems. There are commercially available separation systems that will automatically drain liquid from a separation receiver based on a timed cycle. If liquids have a chance to build up in a vacuum receiver, the probability increases that the water will be re-entrained in the gas stream and eventually contaminate the vacuum pump. It is recommended that the separated liquid be quickly removed from the system.

WATER VAPOR
Now that the removal of liquid water has been discussed, it is time to deal with water vapor inside a vacuum system. Water vapor entrained in the incoming gas stream cannot be effectively separated with a receiver or mechanical separator tank because entrained water vapor is actually part of the gas stream. To deal with vapors properly, there must be a condensation stage so that the water vapor is effectively separated from the incoming gas stream or steps must be taken to ensure that the vapor remains a vapor even after it is compressed back to atmospheric pressure.

The first issue that should be brought up is why do we need to deal with water vapor in the first place. Is it possible to let the water vapor go right through the vacuum pump and then discharge to the ambient atmosphere? With a water sealed vacuum pump, the answer is yes. The liquid seal water helps condense a majority of the incoming water vapor and can actually increase the capacity of the vacuum pump for that particular process. Any water vapor left will exit the vacuum pump with the air stream. When oil sealed vacuum pumps are placed in these types of applications, however, the answer is not always yes. Depending on the amount of saturation and the discharge temperature of the vacuum pump, it may or may not be possible for the vacuum pump to effectively process the incoming water vapor.

To provide an example of how water vapor reacts under vacuum, a cup of water can be used as an example. Assume there is a cup of water, at a temperature of 68° F placed on a table. A clear bell jar is placed over the top of the cup. The bell jar is connected to a vacuum pump and the pump begins to pull vacuum on this system which consists of a bell jar and a 68° F cup of water. Pumping starts at atmospheric pressure (760 torr) and continues on down to 600 torr, then to 500 torr, then to 300 torr, then to 100 torr, all the way down to 17.5 torr. At 17.5 torr a funny thing happens. The water in the cup begins to boil. If the vacuum remains at a constant level, the water will continue to boil and eventually evaporate away (assuming the temperature of the water stays constant). Essentially, what is happening is that as vacuum is pulled on the bell jar, the atmospheric pressure over the surface of the water in the cup is being reduced.

Looking at the dynamics of the system, the boiling water in the cup does no harm to a vacuum pump or system. Actually, if you could put your hand inside the vacuum system and place it in the boiling water, the water would feel cool to the touch even though it is boiling away. The reason it feels cool to the touch is since the pressure of the air surrounding the cup has been reduced, the amount of heat required to boil the water has been reduced as well. Eventually, when deep enough vacuum is drawn on a system, the boiling point of any water trapped inside the system is reached. In essence, the quantity of heat stored in the trapped water is enough to get it to start boiling. In this case (17.5 torr), the quantity of heat is 68° F.

A perfect example of this is food preparation. To boil water at sea level, the temperature of the water has to be increased to 212° F. At elevation, however, where the ambient atmospheric pressure is less, water will boil at lower temperatures. This is why there are different instructions for preparing food at higher elevations than at sea level. Water boiling under vacuum and cooking at altitude are both examples of the same principle.

As was already mentioned, the boiling water itself does no harm to the vacuum pump. All that is happening to the water is that it is changing phase and turning into steam. The steam becomes part of the gas stream inside the bell jar and is pumped away with the other permanent gases. Eventually, the steam and permanent gases are drawn into the vacuum pump. At the inlet of the vacuum pump, the "local" atmospheric conditions are a pressure of 17.5 torr and a temperature of 68° F. Once this air/vapor mixture goes through the vacuum pump, it is compressed back to atmospheric pressure and is eventually discharged from the vacuum pump. The conditions on the discharge side of the vacuum pump, however, are different than the inlet. At the pump discharge port, the pressure is 760 torr and the temperature is whatever the operating discharge temperature is on that particular vacuum pump. When the discharged water vapor sees atmospheric pressure at the backside of the pump, it will immediately condense back to the liquid phase if the temperature is not high enough. A similar process occurs when ambient air is compressed to 100 PSIG with an air compressor. If the temperature is too low or if cooling occurs, the water vapor entrained in the air will immediately condense back to a liquid.

The situation in a vacuum system is similar. Air that is carrying a quantity of water vapor is being drawn in from the vacuum system to the vacuum pump where it is compressed back to atmospheric pressure. If the discharge temperature is too low, liquid water will begin to build up in the lubricating or sealing system. As a result, it will appear that the oil reservoir is filling itself up with oil. What is actually happening is that liquid water is mixing with the lubricating/sealing fluid and giving the appearance of a high oil level.

The dilution of oil with water can have serious effects on a pumping system, especially if the lubricant is used to lubricate pump bearings. Also, the vacuum pump may appear to be pulling less vacuum than before. This is because liquid water entrained in the oil can "flash" back to a vapor in the pump compression chambers. The added water vapor takes up room that would normally be taken up with process air and the pump will appear to have lower capacity.

To determine if a vacuum pump/vacuum application combination will result in a condensation problem, it is necessary to learn a few details about the process. The required pieces of information are the inlet gas stream temperature, the vacuum pump discharge temperature and the operating vacuum level. It can be assumed that the relative humidity in the vacuum system is 100%. In other words, the permanent gases are carrying all the water vapor they can at a particular inlet air temperature. When these three items are known, Figure 8-2 below can be consulted to determine whether or not the vacuum pump will generate condensed water at the discharge port or in the oil reservoir.

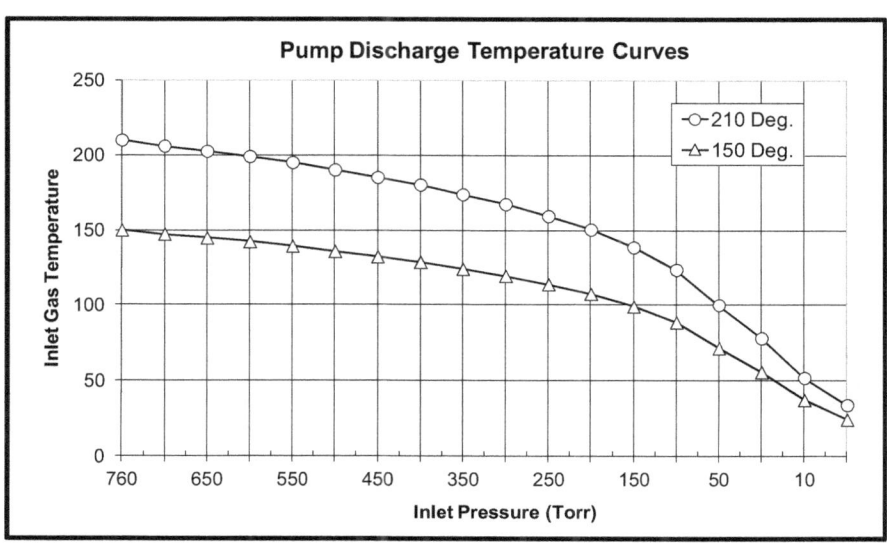

Figure 8-2: Discharge Temperature Condensation Curve

The way to read the chart is to first find the inlet gas stream temperature for the vacuum process. Read the temperature on the Y axis and draw a horizontal line across the page. Then find the operating level of vacuum for the process and read that number on the X axis. Draw a vertical line from that point. Find the location where these two lines meet. The two curves on the graph indicate specific discharge temperatures for an oil sealed vacuum pump. Find the discharge temperature curve for the specific vacuum pump being used and determine if the crossing point on the graph is above or below the curve. If the point is **above** the curve, then condensation will occur under those conditions. If the crossing point is **below** the curve, then condensation will usually not occur.

As an example, take a process that is operating at 50 torr and the incoming gas stream is saturated with water vapor. The pump discharge temperature is 150° F and the incoming gas stream temperature is 100° F. Will there be condensation on the discharge of the vacuum pump? When you find the correct crossover point you find that yes, there will be condensation.

The next question is what to do if there is condensation at the discharge of the vacuum pump. The first course of action is to determine whether it is better to install an inlet condenser or try to keep any vapors in vapor form all the way through the pump. If there is inexpensive or extra chilled water available, the condenser may be the best alternative. If a condenser is used, it is important to then deal with the liquid water that will be generated through condensation of inlet gas stream vapor. See the earlier section on liquid water in vacuum systems.

A different approach may be necessary if a condenser is not the best solution. In some cases, the discharge temperature of the vacuum pump can be raised so that the crossover point on the graph is below the new, higher discharge temperature. **A word of caution about this procedure however**. In oil sealed systems there is the potential for lubricant failure, mechanical failure or even a fire if the discharge temperature is too high for a particular vacuum pump or operating lubricant. It is critical to consult the manufacturer to determine the safe maximum operating temperature. If the discharge temperature can be raised, then it is possible to keep the water vapor in vapor form all the way through the vacuum system. Keep in mind though, that water vapor will condense at the first cool surface it sees, so make certain that discharge piping is positioned so that water does not collect in low points.

Another way to prevent condensation in the vacuum pump is to run a device called a gas ballast. A gas ballast is simply an air inlet valve that injects air into the compression stage of a vacuum pump so that vapors stay in vapor form as they travel through the vacuum pump. Any liquid water entrained in the oil will revert to the vapor phase and be ejected from the pump. The gas ballast valve is built in as standard on many vacuum pumps and will solve some water condensation problems. As a side note, there is sometimes an increase in oil smoke carryover when a gas ballast valve is opened. Make sure there are adequate oil mist separators in place.

Similar in principle to the gas ballast, users can also add a **dry gas purge** to the inlet or discharge reservoir to maintain water vapor in vapor form. Adding a purge to the inlet gas stream requires that extra pumping capacity is available from the vacuum pump. It is not recommended that a small 'leak' is opened on the inlet side to solve a water problem only to create a production process problem. A gas purge will dilute water vapor, reduce the relative humidity in the air stream and help sweep away the vapors before they get a chance to condense. Sometimes, solving some of the worst water condensation problems is as simple as opening a small valve in the inlet piping.

As discussed earlier, start-up and shut-down procedures are important in that vacuum pumps must be up to temperature before process is run. Before shut-down, a dry or clean air purge should be run through the vacuum pump to sweep away any entrained water vapor. It is also very important to have a drip leg installed on the discharge side of vacuum pumps in wet applications so that condensed water in the discharge piping does not drain back into the vacuum pump. See the information in Chapter 6 on oil mist exhaust filtration for recommendations.

Measure the amount of water collected in receivers, separators and discharge lines both before preventative measures are taken and afterward to test the effectiveness of actions to eliminate or reduce water condensation in a vacuum system. Compare the actual results with the targeted results and take additional steps, if necessary. With many oil sealed vacuum pumps, there is also a lubricant analysis program available to test the condition and remaining life on the seal or lubricating fluid. These reports should be checked periodically to ensure that adequate measures are being taken to eliminate the water problem.

Generally, most liquid water and water vapor problems can be solved by taking a close look at vacuum pumping systems and how they relate to inlet gas stream conditions. Many benefits can be realized by taking the necessary measures to eliminate water condensation and liquid contamination from oil sealed vacuum pumps. Vacuum pump life will be increased, maintenance costs will be reduced and most importantly, production run time will be maximized.

Chapter 9

Leaks and Leak Detection

Eventually, every vacuum system will develop a leak. If the leak is of significant size, so as to affect production vacuum levels, it must be located and fixed. One of the most frustrating challenges associated with vacuum systems though, is leak detection. Because vacuum system leaks typically cannot be seen or heard, they can be very difficult to locate, and many times when they are present, other system components are blamed for the problem. When the repair or replacement of these other components does not fix the problem, the trouble shooting process returns to square one.

Outside of these reasons, the primary concern with vacuum leaks is that they add to the production demand load. Leaks can appear to be capacity issues with vacuum pumping systems, when in actuality there is ample available pumping capacity. This chapter briefly discusses the nature of vacuum leaks as well as a few of the common methods for finding them. There are entire books dedicated to the methodology and equipment for locating and fixing leaks in vacuum systems. This chapter is designed only as an introduction to the subject.

Usually, the biggest problem with a vacuum leak is not the actual leak. It is the fact that many production vacuum pumping systems are sized to the very edge of the production demand load. Any additional gas load on the vacuum system from leaks causes an unacceptable reduction in system vacuum level. In essence, what a leak does is add so much SCFM (which then becomes ACFM) flow that the vacuum pump cannot adequately handle it. The result is that process vacuum level suffers. See Figure 9-1 below.

The dotted line represents the SCFM demand flow from the leak and the dashed line represents the normal demand requirements from the production process. The solid line is the total flow requirement for both the leak and production demand flow. As you can see, the closer to atmospheric pressure the process operates, the less effect a leak will have on a vacuum system. This is why many companies utilizing very rough vacuum systems (up near atmospheric pressure) do not have regular leak detection programs in place.

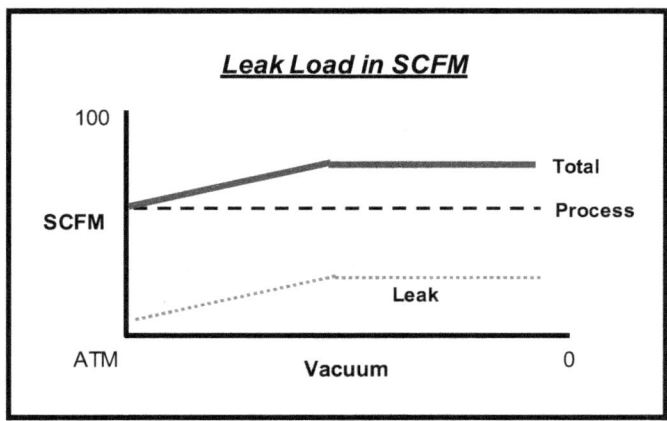

Figure 9-1: Production Load From Leaks

Also note that from about one half of an atmosphere and below, the SCFM flow from the leak remains nearly constant. In most production vacuum systems, the additional load from leaks is less than 5% of the total flow in the system. In many cases it is below 1%. The trick, of course, is to have a system designed and maintained so that the leak load is as small as possible in relation to the pumping capacity of the vacuum system.

There are a two ways to quickly determine the amount of loading that leaks contribute to the overall demand on the vacuum system. For the first method, the specific capacity curve for a vacuum pump must be available. To determine the loading from leaks in a system with this method, first isolate any production machinery from the vacuum system by closing system isolation valves at each machine (if there are multiple machines). Then, operate the smallest possible vacuum pump to get the system as close to its normal operating pressure as possible. If the system cannot reach the normal process level, then go to the next larger vacuum pump. Once the normal system operating vacuum is attained, read the capacity curve for that particular vacuum pump at that level of vacuum. If all the production machinery is isolated from the vacuum system, the only loading on the vacuum pump can be from leaks and the leak load can be determined from the pump capacity curve. This method works for smaller systems but can be cumbersome on larger and more complex systems. This method can be used to get a general idea of how much loading there is from leaks when there are quite a few large leaks in a system.

For smaller leaks, there is another fairly quick method to determine the loading from leaks. For this method, the physical volume of the vacuum system in cubic feet must be known. Once again, isolate production machinery from the vacuum system so that only the supply piping and/or process chamber is open to the vacuum pumping system. In cyclic applications, the process chamber can be checked separately from the supply piping or they can both be checked together. In any case, it is important to have a very accurate reading on the system volume.

Pump the targeted volume down to the lowest typical operating vacuum level and record that number (P_1) in absolute terms (torr, mbar, PSIA, "HgA). After recording P_1, isolate the vacuum pump from the system by closing the inlet isolation valve. Time the pressure rise in the system through the approximate working vacuum range and record these pressures (P_2) at five or ten second intervals. Run this test a couple of times so the results can be averaged to get a more accurate reading of the system. If five or ten second intervals are too short, increase the timed interval to match the pressure rise in the system. When the results of these tests are known, place the values for P_1, P_2, system volume and time into the following formula:

$$S = (V \times (P_2 - P_1) \div t) \times (1 \div P_1)$$

Where:

S = Leak capacity in ACFM
V = Volume of the system in cubic feet
t = Time in minutes
P_1 = Initial pressure
P_2 = Final pressure

When the system variables are put into the formula, the result will be the approximate capacity loading in ACFM from system leaks. Keep in mind that each section of supply piping or portion of the vacuum system can contribute significantly to the loading from leaks. Where possible, it is recommended that each individual section of supply piping or process chamber be checked independently.

PRINCIPLES IN ACTION

A production manager has a hunch that his production vacuum system is leaking at an alarming rate. He pumps the production system down to 12 torr and isolates the vacuum pump from the rest of the system. The rise in pressure to 15 torr takes 7 seconds. What is the ACFM loading from leaks at the base level of vacuum? The volume of the system is 28.3 cubic feet.

SOLUTION:

First, convert the time in seconds to minutes:

$7 \div 60 = 0.12$ **minutes**

Then, place the rest of the values in the formula:

$S = (28.3 \times (15 - 12) \div 0.12) \times (1 \div 12)$

S= 59 ACFM loading from leaks.

Several actions can be taken after determining the level of loading from leaks. If the leak rate is very small, no action may be necessary other then recording the results of the test for a future reference point. If loading is significant, it is important to find out where the leaks are located so repairs can be undertaken. There are four inexpensive methods to determine the location of leaks and several more costly methods.

The easiest and most cost effective method is a soap bubble test. For this type of test, the vacuum system must be pressurized to some reasonable pressure above atmospheric pressure. **A general caution about pressurizing a vacuum system is that most vacuum systems are designed strictly for vacuum service**. Over pressurization may cause individual components to rupture, so it is critical to have an understanding of each system component and design limitations. After the system is pressurized, apply heavily soaped water with a small brush to areas that have a high probability of leakage. Each location where leaks are found should be tagged for future repairs. Depending on the size and number of leaks, this method of leak detection can be used to quickly find many of the larger leaks that are present in a system.

Some vacuum systems cannot be pressurized and must be checked under vacuum. Another simple method for testing for the location of leaks is to use flexible tape on suspected leak areas. The tape will deflect inward when placed over an area with a vacuum leak. This method has been used in many systems where PVC is used for supply piping and the tape is actually left in place as a "fix" for the leak.

The pressure decay test method is used on systems that have isolation valves between individual sections of supply piping. Before using this method to test for and fix leaks, there must be a record of the leak rates for the individual system sections so that comparisons can be made over time. To run this test, the vacuum pumping system is operated and each section of pipe is closed off in sequence from the furthest point on the system to the closest point. All sections must be equipped with vacuum gauges so that the pressure rise can be timed to see how fast it rises from the initial starting point. If the new time for any individual section is significantly faster than the original specification, then a leak is present in that section.

An **ultrasonic leak detector** can be used to locate vacuum leaks by detecting the frequency of the sound of air entering a porosity in a vacuum system. This method will require an investment in new equipment if there is not already one in-house. The frequencies of sound for air entering a leak in a vacuum system can vary widely but can be picked up by commercially available detectors. In operation, the sensor on the leak detector is pointed at the potential leak point. If a leak exists, the leak detector picks up the frequency of air entering the hole or porosity. One problem with this type of leak detection in many installations is the background noise. Much of the sound from vacuum leaks travels inward toward the system and the leak detector may have trouble picking it up if there is a lot of background noise. To solve this problem, there are attachments available to assist with leak detection in vacuum applications. See Appendix B for manufacturers of this type of equipment. Ultrasonic leak detectors are also used in other applications such as locating leaks in compressed air systems, steam traps and in underground pipes. They are very versatile and the cost justification for purchasing one can be spread over several systems. Finding leaks in compressed air systems alone can lead to a very short payback period.

The most accurate leak detector available for rough vacuum systems is the helium leak detector. Helium leak detectors use what is known as a tracer gas to find leaks. The tracer gas, helium in this case, is sprayed on the vacuum system at suspected leak points such as pipe connections, valves and gasketed areas. The sprayed helium is pulled into the vacuum system through the leak path and then travels very rapidly back to the vacuum pumping system. A sensor located in-line at the inlet of the vacuum pump or a sniffer located at the discharge port of the vacuum pump will detect the additional concentration of helium and give an audible signal that a leak is present.

One of the advantages of a helium leak detector is that it can find very small leaks. Helium is a small, light gas and can pass through very small openings in a vacuum system. This same type of leak detection equipment is used in medium and high vacuum applications. The only real downside to the helium leak detector is cost. Some of these systems can be very expensive to purchase and maintain.

The size of the leak will determine the priority in getting it fixed. Many leak detection terms take the form of mass-volume per unit time. In other words, you will see terms like ATM CC's/SEC or MBAR LITERS/SEC when describing leak rates. As a final word on leak detection, it should be stated that a leak detection program should be an on-going event. Waiting until there are production or process problems is too late and can result in system downtime if not carefully managed. There are service companies that specialize in leak detection and by contracting with these companies, the investment in expensive leak detection equipment can be avoided.

Chapter 10

Vacuum Terminology

What compilation of basic industry knowledge would be complete without a description of common industry terms and words? The terms described below are used frequently in the vacuum industry to describe events and conditions relating to vacuum pumps and vacuum applications. When you know the industry buzzwords it is much easier to communicate accurately about conditions surrounding a vacuum pump or vacuum application.

This glossary is intended to be informative so go ahead and read it! In many cases, the solutions to problems in vacuum systems come down to the simplest terms – literally. Understanding the words and terms used in an industry forms the basis for communicating your needs or your clients needs so that problems can be accurately identified and then solved. Intermingled with the vacuum terms are other often used industrial and/or pneumatic terms.

A

Absolute Pressure and Temperature - A term used to describe pressure and temperature that have zero as a base point. Absolute pressures and temperatures are different from gauge pressures and temperatures that use some other point as a reference such as atmospheric pressure or 32 degrees F. Absolute vacuum reads 0″ HgA at perfect vacuum. Absolute temperature is usually measured in Rankine or Kelvin where 0 degrees is the complete absence of heat.

Absolute Rating – A filtration term, it is the size of the largest particle that will pass through a filter element. The diameter of the particle is usually rated in microns.

Absolute Zero - The zero point on an absolute temperature scale (0 degrees Rankine or Kelvin).

Absorb - To draw a material in to another material. Similar to a sponge soaking up water.

Absorption – The embodiment of a gas or vapor in the interior of some other solid.

Accumulation Leak Test – This is a vacuum leak test that is designed to magnify a very small leak. The test gas is collected over time in a closed system or in a sealed part so that a leak rate can be determined.

Accuracy – The degree of precision of a test or instrument. Usually referred to vacuum gauges.

ACFM/CFM – Capacity term in vacuum referring to the volumetric flow rate or actual pumping speed at a given pressure/temperature in cubic feet per minute. Actual Cubic Feet per Minute or Cubic Feet per Minute. ACFM and ICFM have the same meaning.

Activated Carbon – A very highly adsorptive form of carbon. Termed "activated" because it has to be processed to obtain the high adsorptive capacity.

Actuation Pressure - The differential pressure setting of an indicator which signals an actuator or vacuum pump.

Adsorb – The collection of a gas or vapor in condensed form on the surface of a solid material.

Adsorbent - A solid material that adsorbs another material, such as clay, carbon, or activated alumina.

Aerosol – Coalescing term to describe a suspension of very small particles in a gas stream.

Air-Inlet Valve - A valve used for letting atmospheric air into a vacuum system or chamber. Also called a vacuum breaker, air release valve, or vent valve.

Air Regulator - Control device used to meter air into a portion of a vacuum system.

Alternating Current or AC - Electrical supply that reverses direction at a prescribed rate (60 Hz for example). Used in most industrial applications.

Ambient Temperature - The temperature level of the local atmospheric area.

Ambient Pressure - The barometric pressure of the local atmospheric area.

American Institute of Physics - Parent organization of the American Vacuum Society.

American Vacuum Society - An organization that coordinates many of the standards in the vacuum industry.

American Wire Gauge or AWG - A standard used to describe the size of a wire. The larger the value of the AWG number, the thinner the selected wire.

Anti-suck back Valve – Typically, a check valve that is built in to the inlet of a vacuum pump and is designed to prevent the migration of oil and air from a vacuum pump into the system when the vacuum pump stops and the chamber or system is under vacuum. Negates the need for a separate check valve.

Atmospheric Pressure - Pressure exerted by the atmosphere at a specific location. Atmospheric pressure can and does change all the time based on altitude and weather conditions. The standard atmosphere is typically 14.69 PSIA (usually rounded to 14.7) and can also be designated by a mercury column of 760 mm in height at 0 degrees C.

Atomic Mass Unit (AMU) – Used as a basis to measure gases or air in a vacuum system by a residual gas analyzer, the atomic mass unit is the unit of measure of a particle which could be an atom, molecule or ion. The actual value of the particle's mass in AMU is equal to its atomic weight.

Audible Leak Indicator – This is a device that is an accessory to a leak detector. It converts the output leak signal to an audible sound where the frequency is proportional to the relative size of the leak.

B

Backing Pump – A vacuum pump used to evacuate gases from the discharge of another vacuum pump. Can be used for diffusion pumps, turbomolecular pumps, blowers, etc. Also called a forepump.

Background – In leak detection, it is the signal generated by residual tracer gas that is left in the vacuum system. Since the tracer gas cannot be removed quickly, the background signal can build over time. Is also called a "virtual leak" of the tracer gas.

Backstreaming – This is the backwards flow of the oil from a vacuum pump into the chamber or system piping. Usually occurs at higher vacuum conditions when the vapor pressure of the lubricant is reached.

Baffle – Usually a plate or stage that protects a filter or filter elements from direct contact with the contamination entering a system or filter. They are used in many oil and liquid separator designs to keep the final stage free from heavy loading.

Barometric Pressure – The reading of an atmospheric pressure at a certain place and specific time.

Bar – The unit of pressure equal to 106 dynes per square centimeter. This is the European standard atmospheric pressure which is equal to 14.5 PSIA.

Bell Jar - A typically clear container that is open at the bottom, closed at the top and is used as a vacuum chamber or test vessel.

Belt-Drive Vacuum Pump - A pump with the motor drive provided by a belt/sheave system. Sheaves are attached to the pump shaft and the motor shaft. The ratios of the diameters of the pump and motor sheaves determine the actual rotational speed of the pump.

BHP – (Brake Horsepower) This is the total horsepower required at the input shaft of a vacuum pump at specific pressure. BHP curves can be generated to show the input horsepower over the entire vacuum range the pump can achieve.

Blank Off Pressure - The ultimate vacuum that a pump can reach on a closed off inlet port. Also called ultimate pressure of a vacuum pump.

Blind Spots - A location in a filter media where there is no filtration taking place. Can be due to design considerations or other mechanical blocking.

Blinding – A filtration term, it is the reduction or complete shutting off of flow due to closed pores in a filter media.

Booster Pump – Typically a Roots type rotary lobed blower. These pumps act as high capacity "boosters" to mechanical backing pumps at specific vacuum ranges. There are also some vapor pumps that are referred to as booster pumps.

Bridging – Usually seen at seal faces and top plates of filter elements, this refers to particles that arch over individual openings in filter media or through open passages.

Buna-N – Elastomer seals that are made of a Nitrile rubber compound. There are many different formulations of Buna that can be applied to specific applications.

Bypass Valve (Relief Valve) – Typically seen in oil mist exhaust applications, this is a valve mechanism that assures system air flow when a prescribed differential pressure across a filter element is exceeded.

C

Cake – Filtration term referring to solids deposited on a filter media or surface.

Capacity (Dirt Holding Capacity) - The amount of debris or contamination that a filter element is capable of holding without exceeding a specified differential pressure at rated flow.

Cavitation – Applied mainly to water sealed liquid ring vacuum pumps, it is the process where small bubbles are formed inside the compression chambers which then rapidly implode. Usually, this results from too much vacuum applied to a water sealed liquid ring vacuum pump.

Center Tube – The metallic tube inside a filter element designed to support and provide structure to the filter media in a filter element. It also provides adequate fluid flow. The maximum allowable differential pressure determines the structural characteristics of the tube.

Centipoise – A unit of absolute viscosity. One centipoise equals 0.01 poise.

Centistoke - A unit of kinematic viscosity. One centistoke equals 0.01 stoke.

CFR – (Cubic Feet per Revolution) – Used in many rotary lobe blower calculations, it is the volume displaced by a rotary lobed blower in one revolution. CFR will vary with blower size and configuration.

Cleanable Filter Element - A filter element that can be washed, blown out or cleaned when it reaches a given differential pressure. Once the element is cleaned, it must be able to work adequately when placed back in service.

Coalescing – The process where small aerosols and fine mists are combined into larger droplets that can be removed from an air stream. Final coalescing is usually done with a specific coalescing element after the air stream passes through several baffle stages.

Cold Trap or Condenser – An inlet accessory in a vacuum system that is designed to condense vapors prior to the inlet of a vacuum pump. Used in many wet applications and in systems where corrosive or toxic gases are present. Many condenser designs are of the shell and tube configuration.

Collapse Pressure – The pressure at which a filter element deforms to the point where it is allowing bypass of contamination. Also described as the minimum differential pressure that a filter element is designed to withstand without permanent deformation.

Compound Mechanical Pump - A rotary or mechanical vacuum pump having two or more stages of compression in series. Higher vacuum levels are typically attained with more stages of compression.

Compression Ratio - The ratio between the outlet pressure and the inlet pressure of a vacuum pump for a specific gas or gas mixture.

Concentration – Usually expressed in terms like PPM (parts per million) and mg/M^3 (milligrams per cubic meter) it is the amount of one substance entrained in another substance. For example, 1 PPM of oil entrained in an air stream is equal to 0.72 ounces of oil per 100 SCFM and 100 hours.

Conductance – This is the actual capacity of a vacuum piping system in terms of flow. Can be described in ACFM, $M^3/Hour$, Liters/Second, etc. Conductance in a vacuum system can be limited by line size and line configuration.

Contaminant – An unwanted solid, liquid or gas that is present in a vacuum inlet air stream.

Cover Plate – Also called a blank-off plate, it is a cap on an open port in a vacuum system or on the inlet to the vacuum pump.

Cracking – Usually used in terms of vacuum pump oil, it is the decomposition of complex oil molecules into compounds of lower complexity and weight.

Cycle – The process of pulling vacuum on a closed system, performing the desired work, and then venting the system to atmospheric pressure (or other target pressure). Also, it is the length of time a filter assembly is in operation before cleaning is necessary.

Cut In Point – Typically used with vacuum blower systems, it is the pressure that the backing pump has to pull to before the blower can be turned on. After the blower is in operation, the system is pulled down to the target level of vacuum.

D

Degassing – Removing gas from a solid or liquid material under vacuum.

Delta P – Used as a description of component efficiency in a vacuum system, it is the pressure drop or differential pressure across that component.

Density - The mass of a substance for a specific volume of that substance. Expressed in terms such as pounds/cubic foot, or grams/cubic meter.

Differential Pressure – See Delta P. It is the difference in pressure between two points in a vacuum or filter system. In filter assemblies, it is usually expressed as the delta P between a housing inlet and outlet.

Diffusion Pump – A very common high vacuum pump used for industrial applications. A diffusion pump is a vapor pump that has to be operated at high inlet vacuum conditions and must have a mechanical backing pump to compress diffusion pump discharge air back to ATM pressure.

Direct-Drive Pump - A vacuum pump where the electric drive motor is directly coupled to the rotor shaft on the pump. The rotational speed in RPM of the motor is equal to the rotational speed of the vacuum pump.

Displacement – A term used by many vacuum pump manufacturers, it is the geometric swept volume of the compression chamber at typical RPM's. It is also called free air displacement. This value, being theoretical, is not typically used by end-users. It is mainly a used as a standard by vacuum pump manufacturers.

Dissociation - The breakdown of a compound or substance into two or more components. Certain vacuum processes will exhibit dissociation and actually raise pump capacity requirements. Dissociation is also sometimes referred to as cracking.

Drift - The change in the amount of background output in a helium leak detector due to imperfections in the electronics rather than changes in the background level of helium.

E

Effective Filtration Area – The surface area in a filter element that is exposed to process flow and can effectively remove contamination.

Efficiency – In pump terms, it is the ratio of the actual pumping capacity to the theoretical swept volume. In filtration, it is the ability of a filter element to take out contamination. Effi-ciency is usually expressed in terms of a percentage removal rate (i.e. 99.5% efficient at 10 microns).

Equivalent Nitrogen Pressure – Term used by manufacturers and users of electronic vacuum gauges. It is that pressure that a vacuum gauge would read if the system atmosphere was nitrogen instead of the actual system gas mixture.

Explosion-Proof Motor – Used in many process and chemical applications on vacuum pumps, it is an enclosed motor that will withstand an explo-sion of a specific vapor or gas completely within its housing.

F

Factory Calibration - The tuning or altering of a vacuum gauge by the manufacturer to bring (or keep) the device in specification.

Filter Assembly – All parts of an inlet or discharge vacuum filter. Typically consists of a filter housing and filter element. The filter housing will have an inlet port and discharge port and a means to change out the element.

Filter Element – Pleated or wrapped media cartridge that performs the actual filtration process.

Filter Media – The "cloth" used in a filter element. Most vacuum applications use either paper or polyester as the base material.

Filtration - The process of separating solids and vapors from an incoming gas stream.

Flow Rate – The amount of air per unit time that is generated by a production process which then passes through a vacuum pump. Is expressed in volume flow terms such as ACFM or in mass flow terms such as SCFM or Pounds/Hour.

Footprint - The space a piece of equipment occupies on a production work floor or surface.

Foreline - Vacuum line connecting the vacuum system to the inlet of a vacuum pump.

Foreline Valve - A vacuum valve installed in-line so that the backing pump can be isolated from the lead pump or high vacuum pump.

Forepressure - The total pressure on the outlet side of a high vacuum pump measured near the outlet port. Sometimes called the backpressure (or backing pressure), outlet pressure, exhaust pressure, or discharge pressure.

Forepump – See Backing Pump.

Free air – Many times used in error as SCFM, it is air flow at local atmospheric conditions. Temperature, atmospheric pressure and water content are usually not equal to standard conditions so Free Air must be adjusted to account for these differences.

Fuller's Earth – Media used to filter chemical contamination in vacuum pumps. A highly absorbent form of clay.

G

Gas – A gas is a substance in a state of matter where each molecule is not hindered by the forces generated by other molecules (gravity, strong and weak nuclear forces, etc.) so these molecules are free to occupy any space within a container. In vacuum work, a gas can be considered permanent or a vapor which is condensable.

Gas Ballast- A device used to prevent the condensation of vapors in a vacuum pump by admitting a small amount of air into the compression chamber. This device not only prevents the condensation of vapors, it can also help remove condensed vapors in vacuum pump oil.

Gauge Pressure – A vacuum scale that uses atmospheric pressure as a reference point. Inches of Mercury Vacuum for example. On the pressure side, PSIG for example.

GPH - Gallons per hour.

GPM - Gallons per minute.

H

Halogen - The group of elements which includes fluorine, chlorine, bromine, and iodine and similar materials.

Halogen Leak Detector - A leak detector that measures leaks by detecting halogens present in a vacuum system or halogens emanating from a closed, pressurized vacuum system.

Helium Leak Detector - A leak detector that measures leaks by detecting helium present in a vacuum system or helium emanating from a closed, pressurized vacuum system.

High Vacuum - Vacuum range from 0.001 0.000001 torr.

Hot Cathode Ionization Gauge - See Ionization Gauges.

Hydrophilic – A material that attracts water.

Hydrophobic – A material that repels water.

I

ICFM (Inlet Cubic Feet per Minute) – Air flow at inlet conditions to the inlet of rotary lobed blower or booster vacuum pump. In vacuum applications, it is the same as ACFM.

Implosion – Opposite of explosion, it is the result of a vessel or structure not being able to withstand the forces associated with vacuum conditions. The walls of the vessel will rapidly collapse, collide and then rapidly explode outward.

Inches of Water – Units used to measure small pressure differentials across filter components for both vacuum and pressure applications. One inch of water column equals 1.868 torr (or 1 PSI = 27.7″ H_2O).

Inches of Mercury – Two very common scales used to measure vacuum pressures (″HgA and ″HgV). The scale ranges from 29.92″ Hg to 0″ Hg and scale orientation depends on whether it is used as a gauge scale or an absolute scale. One inch of mercury equals 25.4 torr.

Inlet Pressure - The total pressure at the inlet of a vacuum pump.

In-line Type Filter – Used in many blower applications, it is a filter housing where the inlet and outlet ports are located on a common plane or center line.

Ionization Vacuum Gauge - A group of high vacuum gauges that actually ionize gas molecules which are then collected and measured. There are two main types of ion gauges: Hot Cathode and Cold Cathode. Hot cathode ion gauges produce ions by means of a hot filament. This type of gauge is also called a Bayard-Alpert ionization gauge. The Cold Cathode ion gauge (also called the Philips Gauge) produces ions by utilizing a cold cathode discharge in the presence of a magnetic field.

Isolation Test – Sometimes called a pressure decay or rate of rise test, this test measures the amount of leaks present in a vacuum system. The vacuum pump pulls the system down to target operating vacuum level and is then isolated from the system. The rate of pressure rise is timed to determine the leak rate.

Isolation Valve - A valve that seals off a vacuum system from the vacuum pump when the pump is off.

L

L-Type Filter – A filter housing where the inlet and outlet ports are 90 degrees to each other. Vacuum industry standard CSL design.

Leak – A porosity or hole in a vacuum system. Can be detected by a variety of methods such as tracer gas detection, rate of rise testing, ultrasonic detection and soap bubble testing.

Leak Detector – An instrument used to locate leaks in a vacuum system.

Leakage Rate – The amount of air entering a vacuum system. Typically measured in Atmospheric CC's per second or Millibar Liters per Second.

Low Vacuum or Rough Vacuum - Vacuum range from atmospheric pressure to 29.88″ HgV (1 Torr).

M

Mass Number – The molecular weight of a gas or substance (i.e. Air = 29).

Mass Flow – The weight of a gas or air flow going into a vacuum system. Usually expressed in SCFM or Pounds per Hour and is then converted to volume flow (ACFM) for pump sizing.

Mass Spectrometer – In vacuum applications, the two most common Mass Spectrometers are the helium leak detector and the residual gas analyzer. These devices measure gases with specific AMU's to determine either leaks or gases present in vacuum systems.

McLeod Gauge - A liquid level mercury vacuum gauge that measures the pressure in a vacuum system.

Mean Free Path – A value used in many vacuum calculations, it is the average distance that a molecule will travel before it collides with another molecule.

Mechanical Efficiency – Usually expressed as a percent, it is the ratio of the actual pumping speed of a vacuum pump to the theoretical displacement.

Mechanical Pump – General term referring to oil sealed vacuum pumps of liquid ring, rotary vane, rotary screw, rotary piston, etc. design.

Medium Vacuum – The pressure range from 1 torr to 0.001 torr.

Mesh – Filtration term designating the number and size of openings per inch of a material.

Micrometer or Micron - A unit of length. A micron is one millionth of a meter or 0.000039" (39 millionth's of an inch). Expressed in convenient terms, 25 microns is approximately equal one thousandth of an inch (.001").

Micron (Micron of Mercury) – Usually used to measure medium vacuum conditions, it is a unit of pressure equal to 0.001 Torr. Also referred to as a millitorr.

Millibar – One thousandth of a Bar. 1,103 millibar is equal to atmospheric pressure.

Millimeter of Mercury (mmHg) - See Torr.

Millitorr – Pressure measurement equal to 0.001 Torr. See micron.

Molecular Flow – Flow range in vacuum that refers to the random passage of air molecules through a section of pipe or through an orifice. Occurs at high vacuum conditions.

O

OEM - Original Equipment Manufacturer.

Oil Separator Element- Coalescing element designed to trap oil aerosols, mists and droplets prior to the discharge of a vacuum pump.

Open Drip-Proof Motor or ODP - Motor designation that describes an electric motor with ventilator openings that will pre-vent liquids and solids (dropped vertically) from interfer-ing with its operation.

Outgassing - The evolution of gas from a material in a vacuum.

P

Partial Pressure – The pressure exerted by an individual gas in a mixture of gases. For example, an air mixture will exert a total pressure in a system. The constituents that make up air will each exert a portion of that total pressure.
Particle Size Distribution – Filtration term that describes the micron size distribution of a sample of particulate matter from a filter system or vacuum pump.

Perfect Gas — Also called an ideal gas, it is generally a gas that obeys Boyle's and Charles' laws.

Pirani Gauge — A thermal conductivity vacuum gauge used most commonly in rough and medium vacuum applications.

Pounds Per Hour (#/Hr) - A unit of mass flow used in vacuum applications.

Pressure - Force exerted per unit area.

Pressure Differential - The difference of pressure across a component. It is the difference between the pressure on the inlet side of the component and the pressure on the discharge side of a component.

PSID - Pounds per Square Inch Differential, also called Delta P.

PSIG - Pounds per Square Inch Gauge. Uses atmospheric pressure as a reference point.

Pumpdown Curve – A graph that represents the pumpdown of a vacuum system over the course of a specified amount of time. It is used to determine the time required to achieve the target pressure in a vacuum system with a specific vacuum pump.

Pumpdown Time - The amount of time required for a system to pump down from start-up to final operating vacuum levels. Also called the time of evacuation.

Pump Fluid – In rough vacuum, pump fluid refers to the lubricating or seal fluid of a mechanical vacuum pump.

Pumping Speed – Refers to ACFM capacity of a vacuum pump. Also means the amount of gas that can be removed from a system over a period of time.

R

Roughing - The initial pumpdown of a vacuum system.

Roughing Pump – The vacuum pump used to evacuate a high vacuum system to the point where the high vacuum pump can take over.

S

SCFM - Standard Cubic Feet per Minute. The standard for mass flow in vacuum systems at standard conditions: 760 torr, 68 degrees F and 36% RH.

Soap Bubble Test – A method of leak detection where the vacuum system is pressurized with air and suspected leak points are coated with a soapy water solution to determine if leaks are present.

Stage – One compression step in a vacuum system. Vacuum pumps can have several stages directly built in or additional vacuum pumps can serve as addition compression stages (i.e. blower packages).

Staging Ratio - The ratio of the capacity of a booster pump to the capacity of a backing pump.

Standard Room Temperature – There are two values used commonly: 20 degrees C (68 degrees F) or 25 degrees C (77 degrees F).

T

Throughput – A term used frequently to describe flow in medium and high vacuum systems. It is a measure of the amount of gas that passes a cross sectional area of pipe. Throughput units are pressure-volume units – for example Torr-Liters/Second.

Tight (Leak Tight) – Refers to a system with a leak rate that is below a prescribed level. In other words, the leak rate is lower than the system specification.

Time of Evacuation - The time required to pump a given system from atmospheric pressure to a specified base pressure. Also known as pumpdown time or exhaust time.

Tip Speed - **Speed of gear, lobe or helical screw tips expressed in meters per second or in feet per minute.**

Torr – Unit of pressure measurement equal to $1/760^{th}$ of a standard atmosphere. Equal to 1 mm of mercury.

Total Pressure – Refers to the sum of all partial pressures in a gas mixture.

Tracer Gas – The gas, such as helium, that passes through a leak and is then detected helium leak detector. Also termed "search gas".

Transition Flow – One of the three primary flow regimes in vacuum piping, it is the flow of air through a pipe that occurs between viscous flow and molecular flow.

Trap - An accessory used condense vapors present in a vacuum system.

Turbulent Flow – The tumultuous flow of air in a vacuum piping system that occurs at the beginning of viscous flow.

U

Ultimate Pressure - Lowest attainable pressure in a vacuum pump or system. See Blank-Off Pressure.

Ultrahigh Vacuum – Range of vacuum pressures below 1×10^{-6} torr.

Ultrasonic Leak Detector – A device that detects the frequency of the sound of air entering a vacuum system through a hole or porosity.

V

Vacuum – Any pressure in a system that is less than the ambient atmospheric pressure.

Vacuum Cooling – Rapidly evaporating a liquid from the surface of a product or material under vacuum to reduce the temperature. This process is performed regularly on produce prior to transportation.

Vacuum Drying - The removal of a liquid from a substance by evaporation inside a vacuum system. When the liquid is water, the process is sometimes called vacuum dehydration.

Vacuum Gauge - Any instrument used to measure pressure in a vacuum system. Called manometers, diaphragm, thermocouple, Pirani, McLeod, Bourdon Tube, etc.

Vacuum Manifold - Part of a vacuum system piping configuration. Typically, there are a number of branches or ports available so that a number of vacuum processes can be operated simultaneously.

Vacuum System - A pneumatic system designed for the manufacture of a product or the operation of a process. Typically consists of a vacuum pump or pumps, vacuum chamber, interconnecting piping, and a variety of other accessories and components such as filters, gauges and receivers.

Vapor - A material or substance in the gas phase that is condensable at ambient temperatures.

Vapor Pressure - The pressure exerted by the vapor of a liquid when it is in equilibrium with the liquid.

Volumetric Efficiency - The ratio of the actual capacity of a vacuum pump to the theoretical capacity. Typically expressed in percentages.

Virtual Leak - The appearance of a leak in a vacuum system that is caused by the release of trapped gas.

Viscous Flow - One of the three primary flow regimes in vacuum piping, it is the flow of air through a pipe that occurs when the mean free path of the air molecules is very small in relationship to the diameter of the pipe. The flow can be viscous laminar or viscous turbulent.

Chapter 11

Quiz Questions

This section is designed to give the beginner some practice with the concepts and terminology surrounding vacuum technology. The examples here are not real and should be treated as exercises from which the prior learning in this book is put to test. The goal of this book is to give you practice using vacuum terminology as well as practice making basic calculations regarding vacuum applications. Any real life decisions that involve purchasing or installing vacuum equipment should first be reviewed by and then approved by a qualified engineer.

As you gain experience making these types of decisions you will become more confident and eventually be able to do most or all of the work on your own. Until then, use the application examples below to help you get started. The answers are at the end of this chapter.

Pressure Conversions:

1. 25.6″ HgV to HgA = _____

2. 37 Torr to ″HgV = _____

3. 12 PSIA to Torr = _____

4. 2 PSIA to ″HgV = _____

5. 210 millibar to ″HgV = _____

6. 16″ HgV to PSIA = _____

7. 28″ HgV to Millibar = _____

8. 12″ HgA to Millibar = _____

9. 1 Torr to ″HgV = _____

10. 3 Torr to Millibar = _____

11. 19.4″ HgV to PSIA = _____

12. 2 PSIA to Pascal's = _____

13. 723 mmHg to Millibar = _____

14. 89 millibar to ″HgV = _____

15. 89 millibar to PSIA = _____

16. 28.55″ HgV to Pascal's = _____

17. 12″ H_2O to mmHg (gauge) = _____

18. 49 torr to ″HgV = _____

19. 395 millibar to PSIA = _____

20. 12 PSIA to PSIG = _____

21. 50,000 Pascal's to ″HgV = _____

Application Question #1 (Pressure Conversions):

A procurement manager from an international plastics company calls you to ask for help communicating with the company engineers. Apparently, they are all from different countries and all use different vacuum scales. The procurement manager would like to communicate a vacuum specification to all of them in one correspondence and wants to know what 82% of full vacuum is in five other scales. What numbers do you send to the plastics company?

Vacuum Flow Conversions:

1. 35 SCFM @24″ HgV = _____ ACFM

2. 250 SCFM @28.5″ HgV = _____ ACFM

3. 3,500 ACFM @2 Torr = _____ SCFM

4. 1,079 ACFM @15″ HgV = _____ SCFM

5. 7 SCFM @3 Millibar = _____ ACFM

6. 35 ACFM @2″ HgV = _____ SCFM

7. 243 SCFM @2 Torr = _____ ACFM

8. 700 ACFM @12″ HgA = _____ ACFM @4.5″ HgA

9. 35 ACFM @10″ HgV = _____ ACFM @29.9″ HgV

10. 435 Lbs/hour of Air @28″ HgV = _____ ACFM

11. 12 lbs/hour of air @15 Torr = _____ ACFM

13. 300 SCFM at 23° C and 12 mmHg = _____ ACFM

14. 11 SCFM @350° F and 29.9199″ HgV = _____ ACFM

15. 86 SCFM @291 mbar and 33° F = _____ ACFM

16. 700 ACFM at 99° F and 0.015 PSIA = _____ SCFM

17. 2,250 ACFM @528° Rankine and 68 Torr = _____ SCFM

18. 15 ACFM @310° C and 15″ HgA = _____ SCFM

19. 210 ACFM @68° F and 25 Torr = _____ ACFM @90° F and 100 Torr.

20. 55 ACFM @150° C and 10 Mbar = _____ ACFM @490° Rankine and 112 Torr.

Application Question #2 (Flow Conversions):

A maintenance manager from a printing company sends you a specification for a vacuum pump for installation on a printing press. The specification calls for a vacuum pump that will be able to handle 750 SCFM at 18″ HgV and 115° F. What size (ACFM) vacuum pump do you recommend to him?

Application Question #3 (Flow Conversions):

A recent graduate engineer from a chemical company calls you and gives you the following specifications for a vacuum application:

257 Pounds per Hour of Air

175° F.

35 mm Hg

PVC Piping

10 Pounds per Hour of Air Leak Rate

She needs to know what size vacuum pump in ACFM she should use for a mixing operation. What size vacuum pump do you tell her she should use?

Closed System Pumpdown:

1. **System Parameters:**

    Volume = 259 Cubic Feet
    Pumpdown Time = 2 Minutes
    Initial Pressure = ATM
    Desired Pressure = 10 Torr
    Capacity = ?_____

2. **System Parameters:**

    Volume = 17 Cubic Feet
    Pumpdown Time = 20 Seconds
    Initial Pressure = ATM
    Desired Pressure = 28″ HgV
    Capacity = ?_____

3. **System Parameters:**

    Volume = 75 Cubic Feet
    Pumpdown Time = ?_____ Minutes
    Initial Pressure = ATM
    Desired Pressure = 33 Millibar
    Capacity = 17 ACFM

4. **System Parameters:**

Volume = 130 Cubic Feet
Pumpdown Time = ?_____ Minutes
Initial Pressure = ATM
Desired Pressure = 22 Torr
Capacity = 250 ACFM

5. **System Parameters:**

Volume = 13,554 Gallons
Pumpdown Time = 28 Seconds
Initial Pressure = 350 Torr
Desired Pressure = 93 Torr
Capacity = ?_____

6. **System Parameters:**

Volume = 70 Cubic Meters
Pumpdown Time = 2.8 minutes
Initial Pressure = 5″ HgV
Desired Pressure = 29.9″ HgV
Capacity = ?_____

7. <u>System Parameters:</u>

Volume = 3,000 Liters
Pumpdown Time = ?_____ Minutes
Initial Pressure = 750 mmHg
Desired Pressure = 310 Millibar
Capacity = 250 ACFM

8. <u>System Parameters:</u>

Volume = 30,000 Gallons
Pumpdown Time = ?_____ Minutes
Initial Pressure = 760 Torr
Desired Pressure = 0.5 Torr
Capacity = 750 ACFM

9. <u>System Parameters:</u>

Volume = 419 Cubic Feet
Pumpdown Time = 5 Minutes
Initial Pressure = 760 Torr
Desired Pressure = 9 Torr
Capacity = ?_____
Process Flow Rate @9 Torr: 2.2 Standard Liters/Minute

10. <u>System Parameters:</u>

    Volume = 88 Cubic Meters
    Pumpdown Time = 1.5 Minutes
    Initial Pressure = 1013 Millibar
    Desired Pressure = 33 Millibar
    Capacity = ?_____

11. <u>System Parameters:</u>

    Volume = 1,000 Liters
    Pumpdown Time = ?_____ Minutes
    Initial Pressure = 760 Torr
    Desired Pressure = 28 Torr
    Capacity = 860 Cubic Meters/Hour

12. <u>System Parameters:</u>

    Volume = 6 Cubic Feet
    Pumpdown Time = ?_____ Minutes
    Initial Pressure = 0" HgV
    Desired Pressure = 27" HgV
    Capacity = 142 ACFM
    Process Flow Rate @27" HgV: 10 SCFM

Application Question #4 (Pumpdown Calculations):

A vacuum furnace manufacturer requires a vacuum pump to evacuate a vacuum furnace from atmospheric pressure to 28" HgV in 2.2 minutes. The volume of the furnace is 450 cubic feet. The process requires an air flow rate of 50 SCFM at 28" HgV. What size vacuum pump should they use for this application?

Application Question #5 (Pumpdown Calculations):

A hospital lab technician needs to evacuate a small laboratory vacuum chamber and then backfill it with air after they run some tests. The backfill process is done at 16 millibar and requires 28 SCFM of air. The vacuum chamber is 4.9 cubic feet in volume and must be evacuated in 3 minutes or less. What is the correct size vacuum pump to use in this application?

Application Question #6 (Filtration):

A wood working shop is using a 700 ACFM vacuum pump at 24" HgV to hold down plastic sheets so they can cut them into the proper shapes with a CNC cutting machine. They are complaining that the filter replacement intervals are too soon because of the heavy dust loading and want to purchase a filter that gives them longer element life before servicing. How much surface area in media will they need? What is the proper connection for the inlet filter?

Application Question #7 (Filtration):

The above mentioned wood working shop also needs an exhaust demister. What is the proper SCFM flow rating for the filter?

Application Question #8 (Water Vapor):

An application involving saturated inlet air to a vacuum pump could be causing discharge condensation problems. The application involves an oil sealed vacuum pump operating on a drying process. The inlet vacuum level is 225 torr and the inlet gas stream temperature is 100° F. Will a 150° F air/oil discharge temperature on this vacuum pump be high enough to avoid condensation?

Leak Rates:

1. **System Parameters:**

   Volume = 129 Cubic Feet
   Initial Pressure = 50 Torr
   Final Pressure = 75 Torr
   Rise Time = 10 Seconds
   ACFM Leak Rate = ?_____

2. **System Parameters:**

   Volume = 700 Cubic Feet
   Initial Pressure = 250 Torr
   Final Pressure = 350 Torr
   Rise Time = 3 Minutes
   ACFM Leak Rate = ?_____

3. **System Parameters:**

    Volume = 10 Cubic Feet
    Initial Pressure = 100 Torr
    Final Pressure = 160 Torr
    Rise Time = 7 Minutes
    ACFM Leak Rate = ?_____

4. **System Parameters:**

    Volume = 29 Cubic Feet
    Initial Pressure = 1 Torr
    Final Pressure = 3 Torr
    Rise Time = 15 Seconds
    ACFM Leak Rate = ?_____

5. **System Parameters:**

    Volume = 12,00 Liters
    Initial Pressure = 157 torr
    Final Pressure = 300 Torr
    Rise Time = 4.5 Minutes
    ACFM Leak Rate = ?_____

6. **System Parameters:**

   Volume = 45 Cubic Meters
   Initial Pressure = 22″ HgV
   Final Pressure = 19″ HgV
   Rise Time = 13 Minutes
   ACFM Leak Rate = ?_____

7. **System Parameters:**

   Volume = 1,800 Gallons
   Initial Pressure = 30 Millibar
   Final Pressure = 260 Millibar
   Rise Time = 32 Minutes
   ACFM Leak Rate = ?_____

8. **System Parameters:**

   Volume = 3.9 Cubic Feet
   Initial Pressure = 0.5 Torr
   Final Pressure = 1 Torr
   Rise Time = 5 Seconds
   ACFM Leak Rate = ?_____

Application Question #9 (Leak Rate):

A space simulation chamber must be tested for leakage rate. The chamber has a volume of 7.8 liters and the interconnecting piping has a physical volume of 0.5 cubic feet. The system is pumped down to 1 torr and the vacuum pump is isolated from the system. After 23 seconds the pressure rises to 1.25 torr. What is the leak rate in ACFM at 1 torr? What is the SCFM leak rate?

Application Question #10 (Combination):

A can manufacturing plant has a vacuum pump that is down for repair. They would like a rental vacuum pump that can handle the process flow of 388 SCFM at 27″ HgV. Under vacuum the inlet air is heated to 95° F and is saturated with water vapor. They would like an inlet filter, exhaust demister and recommendations on the correct diameter pipe to use in this application. They would like to get a moderate to low amount pressure drop without spending too much money on piping. Also, they would like to operate their vacuum pump with a maximum discharge temperature of 210° F and want to know if they will experience condensation problems. The following information is needed:

1. Correct ACFM size for this application.

2. Inlet particulate filter media surface area.

3. Correct size exhaust demister.

4. Correct Size Pipe.

5. Will they experience condensation?

Application #11 (Combination):

A chemical production facility requires a vacuum pump for a drying application. The application specifications are for 155 pounds per hour of argon at 22 torr and 156° F. The vacuum pump will require an inlet filter for small amounts of particulate matter. The leak specifications on the system call for a chamber with a volume of 900 cubic feet and a pressure rise of not more than 10 torr in 5 minutes. What size vacuum pump in ACFM should be placed in this application and how much filter media surface area should be used?

Now check your answers with the worked out solutions next to see how you did. If there are areas where you incorrectly answered some of the problems, go back and review those sections to make sure you understand those particular concepts completely.

ANSWERS TO QUIZ QUESTIONS

Answers to Pressure Conversions:

1. 25.6" HgV to HgA = **4.32" HgA**

2. 37 Torr to "HgV = **28.46" HgV**

3. 12 PSIA to Torr = **620.5 Torr**

4. 2 PSIA to "HgV = **25.84" HgV**

5. 210 millibar to "HgV = **23.72" HgV**

6. 16" HgV to PSIA = **6.82 PSIA**

7. 28" HgV to Millibar = **65 Millibar**

8. 12" HgA to Millibar = **406.4 Millibar**

9. 1 Torr to "HgV = **29.88" HgV**

10. 3 Torr to Millibar = **4 Millibar**

11. 19.4" HgV to PSIA = **5.2 PSIA**

12. 2 PSIA to Pascal's = **13,790 Pascal's**

13. 723 mmHg to Millibar = **964 Millibar**

14. 89 millibar to "HgV = **27.3" HgV**

15. 89 millibar to PSIA = **1.3 PSIA**

16. 28.55″ HgV to Pascal's = **4,640 Pascal's**

17. 12″ H$_2$O to mmHg (gauge) = **22.4 mmHg**

18. 49 torr to ″HgV = **28″ HgV**

19. 395 millibar to PSIA = **5.73 PSIA**

20. 12 PSIA to PSIG = **-2.7 PSIG**

21. 50,000 Pascal's to ″HgV = **15.2″ HgV**

Answer to Application Question #1 (Pressure Conversions):

A procurement manager from an international plastics company calls you to ask for help communicating with the company engineers. Apparently, they are all from different countries and all use different vacuum scales. The procurement manager would like to communicate a vacuum specification to all of them in one correspondence and wants to know what 82% of full vacuum is in five other scales. What numbers do you send to the plastics company?

82% of Full Vacuum =

136.8 Torr

132.3 Millibar

5.39″ HgA

24.53″ HgV

2.65 PSIA

Answers to Vacuum Flow Conversions:

1. 35 SCFM @24″ HgV = **177 ACFM**

2. 250 SCFM @28.5″ HgV = **5,268 ACFM**

3. 3,500 ACFM @2 Torr = **9.21 SCFM**

4. 1,079 ACFM @15″ HgV = **538 SCFM**

5. 7 SCFM @3 Millibar = **2,364 ACFM**

6. 35 ACFM @2″ HgV = **32.7 SCFM**

7. 243 SCFM @2 Torr = **92,340 ACFM**

8. 700 ACFM @12″ HgA = **1,867 ACFM @4.5″ HgA**

9. 35 ACFM @10″ HgV = **34,860 ACFM @29.9″ HgV**

10. 435 Lbs/hour of Air @28″ HgV = **1,502 ACFM**

11. 12 lbs/hour of air @15 Torr = **134.6 ACFM**

13. 300 SCFM at 23° C and 12 mmHg = **19,194 ACFM**

14. 11 SCFM @350° F and 29.9199″ HgV = **5,049 ACFM**

15. 86 SCFM @291 mbar and 33° F = **279.5 ACFM**

16. 700 ACFM at 99° F and 0.015 PSIA = **0.67 SCFM**

17. 2,250 ACFM @528° Rankine and 68 Torr = **201.3 SCFM**

18. 15 ACFM @310° C and 15″ HgA = **3.78 SCFM**

19. 210 ACFM @68° F & 25 Torr = **54.7 ACFM @90° F, 100 Torr**

20. 55 ACFM @150° C and 10 Mbar = **2.96 ACFM @490° Rankine and 112 Torr.**

Answer to Application Question #2
(Flow Conversions):

A maintenance manager from a printing company sends you a specification for a vacuum pump for installation on a printing press. The specification calls for a vacuum pump that will be able to handle 750 SCFM at 18″ HgV and 115° F. What size (ACFM) vacuum pump do you recommend to him?   **2,050 ACFM**

Answer to Application Question #3
(Flow Conversions):

A recent graduate engineer from a chemical company calls you and gives you the following specifications for a vacuum application:

> 257 Pounds per Hour of Air
> 175° F.
> 35 mm Hg
> PVC Piping
> 10 Pounds per Hour of Air Leak Rate

She needs to know what size vacuum pump in ACFM she should use for a mixing operation. What size vacuum pump do you tell her she should use?   **1,543 Total ACFM (leak rate adds 57.8 ACFM)**

Answers to Closed System Pumpdown:

1. **System Parameters:**
   Volume = 259 Cubic Feet
   Pumpdown Time = 2 Minutes
   Initial Pressure = ATM
   Desired Pressure = 10 Torr
   Capacity = **560.2 ACFM**

2. **System Parameters:**
   Volume = 17 Cubic Feet
   Pumpdown Time = 20 Seconds
   Initial Pressure = ATM
   Desired Pressure = 28″ HgV
   Capacity = **139.9 ACFM**

3. **System Parameters:**
   Volume = 75 Cubic Feet
   Pumpdown Time = **15.1 Minutes**
   Initial Pressure = ATM
   Desired Pressure = 33 Millibar
   Capacity = 17 ACFM

4. **System Parameters:**
   Volume = 130 Cubic Feet
   Pumpdown Time = **1.84 Minutes**
   Initial Pressure = ATM
   Desired Pressure = 22 Torr
   Capacity = 250 ACFM

5. **System Parameters:**
   Volume = 13,554 Gallons
   Pumpdown Time = 28 Seconds
   Initial Pressure = 350 Torr
   Desired Pressure = 93 Torr
   Capacity = **5,141 ACFM**

6. **System Parameters:**
   Volume = 70 Cubic Meters
   Pumpdown Time = 2.8 minutes
   Initial Pressure = 5″ HgV
   Desired Pressure = 29.9″ HgV
   Capacity = **6,285 ACFM**

7. **System Parameters:**
   Volume = 3,000 Liters
   Pumpdown Time = **0.5 Minutes**
   Initial Pressure = 750 mmHg
   Desired Pressure = 310 Millibar
   Capacity = 250 ACFM

8. **System Parameters:**
   Volume = 30,000 Gallons
   Pumpdown Time = **39.1 Minutes**
   Initial Pressure = 760 Torr
   Desired Pressure = 0.5 Torr
   Capacity = 750 ACFM

9. **System Parameters:**
   Volume = 419 Cubic Feet
   Pumpdown Time = 5 Minutes
   Initial Pressure = 760 Torr
   Desired Pressure = 9 Torr
   Capacity = **371.3 ACFM plus 6.6 ACFM = 377.9 ACFM**
   Process Flow Rate @9 Torr: 2.2 Standard Liters/Minute

10. **System Parameters:**
    Volume = 88 Cubic Meters
    Pumpdown Time = 1.5 Minutes
    Initial Pressure = 1013 Millibar
    Desired Pressure = 33 Millibar
    Capacity = **7,085.2 ACFM**

11. **System Parameters:**
    Volume = 1,000 Liters
    Pumpdown Time = **0.23 Minutes**
    Initial Pressure = 760 Torr
    Desired Pressure = 28 Torr
    Capacity = 860 Cubic Meters/Hour

12. **System Parameters:**
    Volume = 6 Cubic Feet
    Pumpdown Time = **0.35 Minutes**
    Initial Pressure = 0″ HgV
    Desired Pressure = 27″ HgV
    Capacity = 142 ACFM
    Process Flow Rate @27″ HgV: 10 SCFM

Answer to Application Question #4 (Pumpdown Calculations):

A vacuum furnace manufacturer requires a vacuum pump to evacuate a vacuum furnace from atmospheric pressure to 28" HgV in 2.2 minutes. The volume of the furnace is 450 cubic feet. The process requires an air flow rate of 50 SCFM at 28" HgV. What size vacuum pump should they use for this application?

561.1 ACFM for Pumpdown and 779.2 ACFM for Process Flow

Answer to Application Question #5 (Pumpdown Calculations):

A hospital lab technician needs to evacuate a small laboratory vacuum chamber and then backfill it with air after they run some tests. The backfill process is done at 16 millibar and requires 28 SCFM of air. The vacuum chamber is 4.9 cubic feet in volume and must be evacuated in 3 minutes or less. What is the correct size vacuum pump to use in this application?

6.8 ACFM for Pumpdown and 1,772 ACFM for Backfill

Answer to Application Question #6 (Filtration):

A wood working shop is using a 700 ACFM vacuum pump at 24" HgV to hold down plastic sheets so they can cut them into the proper shapes with a CNC cutting machine. They are complaining that the filter replacement intervals are too soon because of the heavy dust loading and want to purchase a filter that gives them longer element life before servicing. How much surface area in media will they need? What is the proper connection for the inlet filter?

a. **700/20 = 35 Square Feet of Media for Polyester**
b. **700/7.5 = 93.3 Square Feet of Media for Paper**
c. **5" Pipe Diameter Connection**

Answer to Application Question #7 (Filtration):

The above mentioned wood working shop also needs an exhaust demister. What is the proper SCFM flow rating for the filter?

a. **700 SCFM for Full Flow**
b. **138.5 + 27.7 (20%) = 166.2 for Target Flow Sizing**

Answer to Application Question #8 (Water Vapor):

An application involving saturated inlet air to a vacuum pump could be causing discharge condensation problems. The application involves an oil sealed vacuum pump operating on a drying process. The inlet vacuum level is 225 torr and the inlet gas stream temperature is 100° F. Will a 150° F air/oil discharge temperature on this vacuum pump be high enough to avoid condensation? **Yes**

Answers to Leak Rate Questions:

1. **System Parameters:**
   Volume = 129 Cubic Feet
   Initial Pressure = 50 Torr
   Final Pressure = 75 Torr
   Rise Time = 10 Seconds
   ACFM Leak Rate = **387 ACFM**

2. **System Parameters:**
   Volume = 700 Cubic Feet
   Initial Pressure = 250 Torr
   Final Pressure = 350 Torr
   Rise Time = 3 Minutes
   ACFM Leak Rate = **93.3 ACFM**

3. **System Parameters:**
   Volume = 10 Cubic Feet
   Initial Pressure = 100 Torr
   Final Pressure = 160 Torr
   Rise Time = 7 Minutes
   ACFM Leak Rate = **0.86 ACFM**

4. **System Parameters:**
   Volume = 29 Cubic Feet
   Initial Pressure = 1 Torr
   Final Pressure = 3 Torr
   Rise Time = 15 Seconds
   ACFM Leak Rate = **232 ACFM**

5. **System Parameters:**
   Volume = 12,000 Liters
   Initial Pressure = 157 Torr
   Final Pressure = 300 Torr
   Rise Time = 4.5 Minutes
   ACFM Leak Rate = **85.8 ACFM**

6. **System Parameters:**
   Volume = 45 Cubic Meters
   Initial Pressure = 22" HgV
   Final Pressure = 19" HgV
   Rise Time = 13 Minutes
   ACFM Leak Rate = **46.3 ACFM**

7. **System Parameters:**
    Volume = 1,800 Gallons
    Initial Pressure = 30 Millibar
    Final Pressure = 260 Millibar
    Rise Time = 32 Minutes
    ACFM Leak Rate = **57.7 ACFM**

8. **System Parameters:**
    Volume = 3.9 Cubic Feet
    Initial Pressure = 0.5 Torr
    Final Pressure = 1 Torr
    Rise Time = 5 Seconds
    ACFM Leak Rate = **46.8 ACFM**

Answer to Application Question #9 (Leak Rate):

A space simulation chamber must be tested for leakage rate. The chamber has a volume of 7.8 liters and the interconnecting piping has a physical volume of 0.5 cubic feet. The system is pumped down to 1 torr and the vacuum pump is isolated from the system. After 23 seconds the pressure rises to 1.25 torr. What is the leak rate in ACFM at 1 torr? What is the SCFM leak rate?

    a. **ACFM = 0.51**
    b. **SCFM = 0.00067**

Answer to Application Question #10 (Combination):

A can manufacturing plant has a vacuum pump that is down for repair. They would like a rental vacuum pump that can handle the process flow of 388 SCFM at 27" HgV. Under vacuum the inlet air is heated to 95° F and is saturated with water vapor. They would like an inlet filter, exhaust demister and recommendations on the correct diameter pipe to use in this application. They would like to get a moderate to low amount pressure drop without spending too much money on piping. Also, they would like to operate their vacuum pump with a maximum discharge temperature of $210°$ F and want to know if they will experience condensation problems. The following information is needed:

1. Correct ACFM size for this application.

2. Inlet particulate filter media surface area.

3. Correct size exhaust demister.

4. Correct Size Pipe.

5. Will they experience condensation?

1. **4,179 ACFM**

2. **119.4 Square Feet**

3. **4,179 SCFM Exhaust Demister (Full Flow)**

4. **12″ Diameter Pipe**

5. **No Condensation**

Answer to Application #11 (Combination):

A chemical production facility requires a vacuum pump for a drying application. The application specifications are for 155 pounds per hour of argon at 22 torr and $156°$ F. The vacuum pump will require an inlet filter for small amounts of particulate matter. The leak specifications on the system call for a chamber with a volume of 900 cubic feet and a pressure rise of not more than 10 torr in 5 minutes. What size vacuum pump in ACFM should be placed in this application and how much filter media surface area should be used?

a. **1,002 ACFM (Pump) + 82 ACFM (Leaks) = 1,084 ACFM**
b. **31 Square Feet @ 35 Feet/Minute Face Velocity**

The Final Word

Thank you for purchasing this text! Through reading and studying the material in this book, you have developed new skills and have made yourself more profitable (or marketable) by spending the time to learn what was in these pages.

You have made quite a large step toward a self-sufficient knowledge of basic vacuum technology. Learning the terminology and concepts presented here should provide you with the confidence to communicate effectively about topics in and around vacuum applications. There are many excellent reference books and articles where you can continue to learn about and work with vacuum technology. Try some of these materials to hone your skills with vacuum. Good Luck!

Appendix A:
Providers of Vacuum Related Equipment

Vacuum Filtration Products

Solberg Manufacturing Inc.
1151 W. Ardmore
Itasca, IL 60143
(630) 773-1363 – Phone
(630) 773-0727 – Fax
Email: **sales@solbergmfg.com**

Lubricants and Lubricant Analysis

CPI Engineering Services Inc.
 2300 James Savage Road
Midland, MI 48642
(989) 496-3780 – Phone
(989) 496-2313 – Fax
www.cpieng.com

Particulate Analysis/Particle Distribution

Particle Technology Labs, LTD
555 Rogers Street
Downers Grove, IL 60515
(630) 969-2703 – Phone
(630) 969-2745 – Fax
www.particletechlabs.com

Ultra Sonic Leak Detectors

U.E. Systems
14 Hayes Street
Elmsford, NY 10523
(914) 592-1220 – Phone
(914) 347-2181 – Fax
info@uesystems.com

SDT North America
PO Box 682
Cobourg, ON - K9A 4R5
Canada
(800) 667-5325
(800) 224-1546
info@SDTHearMore.com

Appendix B: Conversion Tables

From	Multiply By:	To Get:
Atmospheres	1.013	Bar
Atmospheres	76	Centimeters of Mercury
Atmospheres	33.9	Feet of Water
Atmospheres	29.92	Inches of Mercury
Atmospheres	406.8	Inches of Water
Atmospheres	1.0332	Kilograms/Square Centimeter
Atmospheres	10,332	Kilograms/Square Meter
Atmospheres	101.325	Kilopascals
Atmospheres	760,000	Microns
Atmospheres	1013	Millibar
Atmospheres	10,332	Millimeters of Water
Atmospheres	101,325	Pascals
Atmospheres	2,116.2	Pounds/Square Foot
Atmospheres	14.7	Pounds/Square Inch
Atmospheres	760	Torr (Millimeters of Mercury)
Bar	0.9869	Atmospheres
Bar	75	Centimeters of Mercury
Bar	33.46	Feet of Water
Bar	29.528	Inches of Mercury
Bar	401.48	Inches of Water
Bar	1.02	Kilograms/Square Centimeter
Bar	10,196.9	Kilograms/Square Meter
Bar	100	Kilopascals
Bar	750,000	Microns
Bar	1,000	Millibar
Bar	10,196.9	Millimeters of Water
Bar	100,000	Pascals
Bar	2,088.5	Pounds/Square Foot
Bar	14.5	Pounds/Square Inch
Bar	750	Torr (Millimeters of Mercury)
cc's/Minute	0.01667	cc's/Second
cc's/Minute	0.00003531	Cubic Feet/Minute
cc's/Minute	0.000006	Cubic Meters/Hour
cc's/Minute	0.000001	Cubic Meters/Minute
cc's/Minute	0.00000001667	Cubic Meters/Second
cc's/Minute	0.000264	Gallons/Minute
cc's/Minute	0.001	Liters/Minute
cc's/Minute	0.00001667	Liters/Second

From	Multiply By:	To Get:
cc's/Second	60	cc's/Minute
cc's/Second	0.00212	Cubic Feet/Minute
cc's/Second	0.0036	Cubic Meters/Hour
cc's/Second	0.00006	Cubic Meters/Minute
cc's/Second	0.000001	Cubic Meters/Second
cc's/Second	0.01585	Gallons/Minute
cc's/Second	0.06	Liters/Minute
cc's/Second	0.001	Liters/Second
Centigrade	(°C x 9/5) + 32	Fahrenheit
Centigrade	°C + 273.18	Kelvin
Cent. of Mercury	0.01316	Atmospheres
Cent. of Mercury	0.01333	Bar
Cent. of Mercury	0.4461	Feet of Water
Cent. of Mercury	0.3937	Inches of Mercury
Cent. of Mercury	5.35	Inches of Water
Cent. of Mercury	0.01359	Kilograms/Square Centimeter
Cent. of Mercury	135.95	Kilograms/Square Meter
Cent. of Mercury	1.333	Kilopascals
Cent. of Mercury	10,000	Microns
Cent. of Mercury	13.33	Millibar
Cent. of Mercury	135.95	Millimeters of Water
Cent. of Mercury	1,333.22	Pascals
Cent. of Mercury	27.85	Pounds/Square Foot
Cent. of Mercury	0.1934	Pounds/Square Inch
Cent. of Mercury	10	Torr (Millimeters of Mercury)
Cubic Centimeters	0.00003531	Cubic Feet
Cubic Centimeters	0.06102	Cubic Inches
Cubic Centimeters	0.000001	Cubic Meters
Cubic Centimeters	0.000001308	Cubic Yards
Cubic Centimeters	0.0002642	Gallons
Cubic Centimeters	0.001	Liters
Cubic Centimeters	0.033814	Ounces (fluid)
Cubic Centimeters	0.002113	Pints
Cubic Centimeters	0.001057	Quarts
Cubic Feet	28,316.9	Cubic Centimeters
Cubic Feet	1,728	Cubic Inches
Cubic Feet	0.02832	Cubic Meters
Cubic Feet	0.03704	Cubic Yards

Cubic Feet	7.48	Gallons
Cubic Feet	28.32	Liters
Cubic Feet	957.51	Ounces (fluid)
Cubic Feet	59.84	Pints
Cubic Feet	29.92	Quarts
Cubic Feet/Minute	28,317	cc's/Minute
Cubic Feet/Minute	471.95	cc's/Second
Cubic Feet/Minute	1.699	Cubic Meters/Hour
Cubic Feet/Minute	0.02832	Cubic Meters/Minute
Cubic Feet/Minute	0.000472	Cubic Meters/Second
Cubic Feet/Minute	7.48	Gallons/Minute
Cubic Feet/Minute	28.32	Liters/Minute
Cubic Feet/Minute	0.472	Liters/Second
Cubic Inches	16.39	Cubic Centimeters
Cubic Inches	0.0005787	Cubic Feet
Cubic Inches	0.00001639	Cubic Meters
Cubic Inches	0.00002143	Cubic Yards
Cubic Inches	0.004329	Gallons
Cubic Inches	0.01639	Liters
Cubic Inches	0.5541	Ounces (fluid)
Cubic Inches	0.03463	Pints
Cubic Inches	0.01732	Quarts
Cubic Meters	1,000,000	Cubic Centimeters
Cubic Meters	35.31	Cubic Feet
Cubic Meters	61,023.7	Cubic Inches
Cubic Meters	1.308	Cubic Yards
Cubic Meters	264.17	Gallons
Cubic Meters	1,000	Liters
Cubic Meters	33,814	Ounces (fluid)
Cubic Meters	2,113.38	Pints
Cubic Meters	1,056.7	Quarts
Cubic Meters/Hour	16,667	cc's/Minute
Cubic Meters/Hour	277.8	cc's/Second
Cubic Meters/Hour	0.589	Cubic Feet/Minute
Cubic Meters/Hour	0.01667	Cubic Meters/Minute
Cubic Meters/Hour	0.000278	Cubic Meters/Second
Cubic Meters/Hour	4.403	Gallons/Minute
Cubic Meters/Hour	16.667	Liters/Minute

From	Multiply By:	To Get:
Cubic Meters/Hour	0.2778	Liters/Second
Cubic Meters/Minute	1,000,000	cc's/Minute
Cubic Meters/Minute	16,667	cc's/Second
Cubic Meters/Minute	35.31	Cubic Feet/Minute
Cubic Meters/Minute	60	Cubic Meters/Hour
Cubic Meters/Minute	0.01667	Cubic Meters/Second
Cubic Meters/Minute	264.17	Gallons/Minute
Cubic Meters/Minute	1,000	Liters/Minute
Cubic Meters/Minute	60,000	Liters/Second
Cubic Meters/Second	60,000,000	cc's/Minute
Cubic Meters/Second	1,000,000	cc's/Second
Cubic Meters/Second	2,118.88	Cubic Feet/Minute
Cubic Meters/Second	3,600	Cubic Meters/Hour
Cubic Meters/Second	60	Cubic Meters/Minute
Cubic Meters/Second	15,900	Gallons/Minute
Cubic Meters/Second	60,000	Liters/Minute
Cubic Meters/Second	1,000	Liters/Second
Cubic Yards	764,555	Cubic Centimeters
Cubic Yards	27	Cubic Feet
Cubic Yards	46,656	Cubic Inches
Cubic Yards	0.7646	Cubic Meters
Cubic Yards	201.97	Gallons
Cubic Yards	764.6	Liters
Cubic Yards	25,852.7	Ounces (fluid)
Cubic Yards	1,615.8	Pints
Cubic Yards	807.9	Quarts
Fahrenheit	($^{\circ}$F - 32) x 5/9	Centigrade
Fahrenheit	$^{\circ}$F + 460	Rankine
Gallons	3,785.4	Cubic Centimeters
Gallons	0.1337	Cubic Feet
Gallons	231	Cubic Inches
Gallons	0.003785	Cubic Meters
Gallons	0.004951	Cubic Yards
Gallons	3.79	Liters
Gallons	128	Ounces (fluid)
Gallons	8	Pints
Gallons	4	Quarts
Gallons/Minute	3,785.4	cc's/Minute

From	Multiply By:	To Get:
Gallons/Minute	63.1	cc's/Second
Gallons/Minute	0.1337	Cubic Feet/Minute
Gallons/Minute	0.2271	Cubic Meters/Hour
Gallons/Minute	0.00379	Cubic Meters/Minute
Gallons/Minute	0.00006309	Cubic Meters/Second
Gallons/Minute	3.785	Liters/Minute
Gallons/Minute	0.0631	Liters/Second
Inches of Mercury	0.03342	Atmospheres
Inches of Mercury	0.03387	Bar
Inches of Mercury	2.54	Centimeters of Mercury
Inches of Mercury	1.133	Feet of Water
Inches of Mercury	13.6	Inches of Water
Inches of Mercury	0.0345	Kilograms/Square Centimeter
Inches of Mercury	345.32	Kilograms/Square Meter
Inches of Mercury	3.387	Kilopascals
Inches of Mercury	25,400	Microns
Inches of Mercury	33.87	Millibar
Inches of Mercury	345.32	Millimeters of Water
Inches of Mercury	3,386.5	Pascals
Inches of Mercury	70.73	Pounds/Square Foot
Inches of Mercury	0.4912	Pounds/Square Inch
Inches of Mercury	25.4	Torr (Millimeters of Mercury)
Inches of Water	0.002458	Atmospheres
Inches of Water	0.002491	Bar
Inches of Water	0.01868	Centimeters of Mercury
Inches of Water	0.0833	Feet of Water
Inches of Water	0.07355	Inches of Mercury
Inches of Water	0.00254	Kilograms/Square Centimeter
Inches of Water	25.4	Kilograms/Square Meter
Inches of Water	0.2491	Kilopascals
Inches of Water	1,868.2	Microns
Inches of Water	2.491	Millibar
Inches of Water	25.4	Millimeters of Water
Inches of Water	249.08	Pascals
Inches of Water	5.202	Pounds/Square Foot
Inches of Water	0.03613	Pounds/Square Inch
Inches of Water	1.868	Torr (Millimeters of Mercury)
Kelvin	$^{\circ}K \times 1.8$	Rankine

From	Multiply By:	To Get:
Kilograms/Sq. Cent.	0.9679	Atmospheres
Kilograms/Sq. Cent.	0.9807	Bar
Kilograms/Sq. Cent.	73.56	Centimeters of Mercury
Kilograms/Sq. Cent.	32.81	Feet of Water
Kilograms/Sq. Cent.	28.96	Inches of Mercury
Kilograms/Sq. Cent.	393.7	Inches of Water
Kilograms/Sq. Cent.	10,000	Kilograms/Square Meter
Kilograms/Sq. Cent.	98.07	Kilopascals
Kilograms/Sq. Cent.	735,579	Microns
Kilograms/Sq. Cent.	980.7	Millibar
Kilograms/Sq. Cent.	10,000	Millimeters of Water
Kilograms/Sq. Cent.	98,069	Pascals
Kilograms/Sq. Cent.	2048.2	Pounds/Square Foot
Kilograms/Sq. Cent.	14.22	Pounds/Square Inch
Kilograms/Sq. Cent.	735.6	Torr (Millimeters of Mercury)
Kilograms/Sq. Meter	0.00009679	Atmospheres
Kilograms/Sq. Meter	0.00009807	Bar
Kilograms/Sq. Meter	0.007356	Centimeters of Mercury
Kilograms/Sq. Meter	0.003281	Feet of Water
Kilograms/Sq. Meter	0.002896	Inches of Mercury
Kilograms/Sq. Meter	0.03937	Inches of Water
Kilograms/Sq. Meter	0.0001	Kilograms/Square Centimeter
Kilograms/Sq. Meter	73.56	Microns
Kilograms/Sq. Meter	0.09804	Millibar
Kilograms/Sq. Meter	1	Millimeters of Water
Kilograms/Sq. Meter	9.807	Pascals
Kilograms/Sq. Meter	0.2048	Pounds/Square Foot
Kilograms/Sq. Meter	0.001422	Pounds/Square Inch
Kilograms/Sq. Meter	0.07356	Torr (Millimeters of Mercury)
Kilopascals	0.009869	Atmospheres
Kilopascals	0.01	Bar
Kilopascals	0.75	Centimeters of Mercury
Kilopascals	0.335	Feet of Water
Kilopascals	0.2953	Inches of Mercury
Kilopascals	4.015	Inches of Water
Kilopascals	0.0102	Kilograms/Square Centimeter
Kilopascals	101.97	Kilograms/Square Meter
Kilopascals	7,500.6	Microns

From	Multiply By:	To Get:
Kilopascals	10	Millibar
Kilopascals	101.97	Millimeters of Water
Kilopascals	1,000	Pascals
Kilopascals	20.89	Pounds/Square Foot
Kilopascals	0.14504	Pounds/Square Inch
Kilopascals	7.5	Torr (Millimeters of Mercury)
Liters	1,000	Cubic Centimeters
Liters	0.03531	Cubic Feet
Liters	61.02	Cubic Inches
Liters	0.001	Cubic Meters
Liters	0.001308	Cubic Yards
Liters	0.2642	Gallons
Liters	33.81	Ounces (fluid)
Liters	2.11	Pints
Liters	1.057	Quarts
Liters/Minute	1,000	cc's/Minute
Liters/Minute	16.67	cc's/Second
Liters/Minute	0.03531	Cubic Feet/Minute
Liters/Minute	0.06	Cubic Meters/Hour
Liters/Minute	0.001	Cubic Meters/Minute
Liters/Minute	0.00001667	Cubic Meters/Second
Liters/Minute	0.2642	Gallons/Minute
Liters/Minute	0.01667	Liters/Second
Liters/Second	60,000	cc's/Minute
Liters/Second	1,000	cc's/Second
Liters/Second	2.119	Cubic Feet/Minute
Liters/Second	3.6	Cubic Meters/Hour
Liters/Second	0.06	Cubic Meters/Minute
Liters/Second	0.001	Cubic Meters/Second
Liters/Second	15.85	Gallons/Minute
Liters/Second	60	Liters/Minute
Microns	0.000001316	Atmospheres
Microns	0.000001333	Bar
Microns	0.0001	Centimeters of Mercury
Microns	0.00004461	Feet of Water
Microns	0.00003937	Inches of Mercury
Microns	0.0005353	Inches of Water
Microns	0.000001359	Kilograms/Square Centimeter

From	Multiply By:	To Get:
Microns	0.01359	Kilograms/Square Meter
Microns	0.0001333	Kilopascals
Microns	0.001333	Millibar
Microns	0.01359	Millimeters of Water
Microns	0.13333	Pascals
Microns	0.002784	Pounds/Square Foot
Microns	0.00001934	Pounds/Square Inch
Microns	0.001	Torr (Millimeters of Mercury)
Millibar	0.0009869	Atmospheres
Millibar	0.001	Bar
Millibar	0.075	Centimeters of Mercury
Millibar	0.03346	Feet of Water
Millibar	0.02953	Inches of Mercury
Millibar	0.4015	Inches of Water
Millibar	0.00102	Kilograms/Square Centimeter
Millibar	10.2	Kilograms/Square Meter
Millibar	0.1	Kilopascals
Millibar	750	Microns
Millibar	10.2	Millimeters of Water
Millibar	100	Pascals
Millibar	2.089	Pounds/Square Foot
Millibar	0.0145	Pounds/Square Inch
Millibar	0.7501	Torr (Millimeters of Mercury)
Millimeters	0.1	Centimeters
Millimeters	0.003281	Feet
Millimeters	0.03937	Inches
Millimeters	0.001	Meters
Millimeters	1,000	Microns
Millimeters	0.001094	Yards
Millimeters of Water	0.00009678	Atmospheres
Millimeters of Water	0.00009806	Bar
Millimeters of Water	0.007355	Centimeters of Mercury
Millimeters of Water	0.00328	Feet of Water
Millimeters of Water	0.002896	Inches of Mercury
Millimeters of Water	0.03337	Inches of Water
Millimeters of Water	0.0001	Kilograms/Square Centimeter
Millimeters of Water	1	Kilograms/Square Meter
Millimeters of Water	0.009806	Kilopascals

From	Multiply By:	To Get:
Millimeters of Water	73.55	Microns
Millimeters of Water	0.09806	Millibar
Millimeters of Water	9.806	Pascals
Millimeters of Water	0.2048	Pounds/Square Foot
Millimeters of Water	0.001422	Pounds/Square Inch
Millimeters of Water	0.07355	Torr (Millimeters of Mercury)
Ounces (fluid)	29.57	Cubic Centimeters
Ounces (fluid)	0.001044	Cubic Feet
Ounces (fluid)	1.805	Cubic Inches
Ounces (fluid)	0.00002957	Cubic Meters
Ounces (fluid)	0.00003868	Cubic Yards
Ounces (fluid)	0.007813	Gallons
Ounces (fluid)	0.02957	Liters
Ounces (fluid)	0.0625	Pints
Ounces (fluid)	0.03125	Quarts
Pascals	0.000009869	Atmospheres
Pascals	0.00001	Bar
Pascals	0.0007501	Centimeters of Mercury
Pascals	0.0003346	Feet of Water
Pascals	0.0002953	Inches of Mercury
Pascals	0.004015	Inches of Water
Pascals	0.0000102	Kilograms/Square Centimeter
Pascals	0.102	Kilograms/Square Meter
Pascals	0.001	Kilopascals
Pascals	7.5	Microns
Pascals	0.01	Millibar
Pascals	0.102	Millimeters of Water
Pascals	1	Newtons/Square Meter
Pascals	0.0209	Pounds/Square Foot
Pascals	0.000145	Pounds/Square Inch
Pascals	0.007501	Torr (Millimeters of Mercury)
Pints	473.18	Cubic Centimeters
Pints	0.01671	Cubic Feet
Pints	28.875	Cubic Inches
Pints	0.0004732	Cubic Meters
Pints	0.0006189	Cubic Yards
Pints	0.125	Gallons
Pints	0.473	Liters

From	Multiply By:	To Get:
Pints	16	Ounces (fluid)
Pints	0.5	Quarts
Pounds/Square Foot	0.0004725	Atmospheres
Pounds/Square Foot	0.0004788	Bar
Pounds/Square Foot	0.03591	Centimeters of Mercury
Pounds/Square Foot	0.01602	Feet of Water
Pounds/Square Foot	0.01414	Inches of Mercury
Pounds/Square Foot	0.1922	Inches of Water
Pounds/Square Foot	0.0004882	Kilograms/Square Centimeter
Pounds/Square Foot	4.882	Kilograms/Square Meter
Pounds/Square Foot	0.04788	Kilopascals
Pounds/Square Foot	359.13	Microns
Pounds/Square Foot	4.882	Millimeters of Water
Pounds/Square Foot	47.88	Pascals
Pounds/Square Foot	0.006944	Pounds/Square Inch
Pounds/Square Foot	0.3591	Torr (Millimeters of Mercury)
Pounds/Square Inch	0.06805	Atmospheres
Pounds/Square Inch	0.06895	Bar
Pounds/Square Inch	5.171	Centimeters of Mercury
Pounds/Square Inch	2.31	Feet of Water
Pounds/Square Inch	2.036	Inches of Mercury
Pounds/Square Inch	27.68	Inches of Water
Pounds/Square Inch	0.07031	Kilograms/Square Centimeter
Pounds/Square Inch	703.1	Kilograms/Square Meter
Pounds/Square Inch	6.895	Kilopascals
Pounds/Square Inch	51,715.1	Microns
Pounds/Square Inch	68.95	Millibar
Pounds/Square Inch	703.1	Millimeters of Water
Pounds/Square Inch	6,894.8	Pascals
Pounds/Square Inch	144	Pounds/Square Foot
Pounds/Square Inch	51.71	Torr (Millimeters of Mercury)
Quarts	946.35	Cubic Centimeters
Quarts	0.03342	Cubic Feet
Quarts	57.75	Cubic Inches
Quarts	0.0009464	Cubic Meters
Quarts	0.001238	Cubic Yards
Quarts	0.25	Gallons
Quarts	0.9464	Liters

From	Multiply By:	To Get:
Quarts	32	Ounces (fluid)
Quarts	2	Pints
Rankine	°R - 460	Fahrenheit
Rankine	°R x 0.5556	Kelvin
Torr (mm of Mercury)	0.001316	Atmospheres
Torr (mm of Mercury)	0.001333	Bar
Torr (mm of Mercury)	0.1	Centimeters of Mercury
Torr (mm of Mercury)	0.04461	Feet of Water
Torr (mm of Mercury)	0.03937	Inches of Mercury
Torr (mm of Mercury)	0.05353	Inches of Water
Torr (mm of Mercury)	0.001359	Kilograms/Square Centimeter
Torr (mm of Mercury)	13.6	Kilograms/Square Meter
Torr (mm of Mercury)	0.1333	Kilopascals
Torr (mm of Mercury)	1,000	Microns
Torr (mm of Mercury)	1.333	Millibar
Torr (mm of Mercury)	1	Millimeters of Mercury
Torr (mm of Mercury)	13.6	Millimeters of Water
Torr (mm of Mercury)	133.32	Pascals
Torr (mm of Mercury)	2.784	Pounds/Square Foot
Torr (mm of Mercury)	0.01934	Pounds/Square Inch

Appendix C:

Useful Vacuum Formulas

SCFM to ACFM:

$\text{SCFM} \times (P_1 \div P_2) = \text{ACFM}$

ACFM to SCFM:

$\text{ACFM} \times (P_2 \div P_1) = \text{SCFM}$

SCFM to ACFM (w/temperature):

$\text{SCFM} \times (P_1 \div P_2) \times (T_2 \div T_1) = \text{ACFM}$

ACFM to ACFM:

$\text{ACFM} \times (P_1 \div P_2) = \text{ACFM (adjusted)}$

Pumpdown of a Closed System:

$S = 2.3 \times (V \div T) \times \log(P_1 \div P_2)$ Solve for Capacity

$T = 2.3 \times (V \div S) \times \log(P_1 \div P_2)$ Solve for Time

Determine Receiver Size (V2):

$V2 = V1 \times (P_1 - P_3) \div (P_3 - P_2)$

Solve Leak Rate for Capacity:

$S = (V \times (P_2 - P_1) \div t) \times (1 \div P_1)$

Useful Vacuum Formulas

Pounds per Hour to ACFM:

$S = (\text{lbs.} \div 60) \times (385 \div MW) \times (760 \div P_2) \times ((460 + T) \div 528)$

Surface Area of a Pipe:

$3.1416 \times R \times R$ (radius should be in feet)

Face Velocity:

F.V. (in feet/min) = ACFM or SCFM ÷ Square Feet of Media

Degrees Fahrenheit to Degrees Rankine:

Degrees F + 460 = Degrees R

Appendix D:
Vacuum System Auditing

Save energy, lower operating costs and optimize vacuum delivery to use points with a vacuum system audit. A vacuum system audit typically consists of the following steps:

Step #1 – In-Briefing Meeting

The first step is a one hour meeting with anyone involved in making decisions about the vacuum systems. This includes personnel from engineering, production, accounting, facilities management and general management. This meeting is very important because it is where we set the site priorities for the vacuum system and it provides us with information on how to match our process with site goals and requirements.

Step #2 – Begin On-Site Audit

Next is the on-site evaluation of your vacuum systems including supply equipment, distribution systems and demand applications. At this time we install electronic data loggers that record current and vacuum at 8 second intervals. Data loggers are left on the system for the duration of the audit and are downloaded each day so that system performance can be evaluated in the evenings. For some applications, high speed data loggers that record at intervals of 25 times per second are used to capture very fast point-of-use events.

Step #3 – Out-Briefing Meeting

At the conclusion of the on-site audit, we would like to meet again with key site personnel to report on our initial findings and to discuss strategies for meeting your vacuum system goals. This meeting provides a final opportunity to us to answer any questions regarding our findings, conclusions and initial recommendations while we are still on-site.

Step #4 – Report Generation

We will spend the next several days analyzing data logger graphs and other information that we gathered while on-site. A report is generated that includes specific narrative regarding your vacuum systems and production equipment along with supporting information in the Attachments portion of the report.

Step #5 – Deliver the Audit Report

Your vacuum system audit report will be delivered either in person or electronically and will be in Microsoft Word or Adobe PDF format. We will be happy to go over the findings and recommendations with you either in person or via teleconference.

A typical audit report will contain the following information and key deliverables:

- **General Observations**
- **Existing System Operation and Layout**
- **Proposed System Operation and Layout**
- **Existing Costs and Proposed Costs**
- **Action Plans for System Improvements**

*Examples of Key Deliverables – Next 2 Pages

Plastics Manufacturing Company
Mobile, AL

Energy Calculations						
Existing Arrangement	**bhp**	**kW**	**Hours**	**KWH**	**Costs**	
Peak Production	1,131	903	3,650	3,295,725	$	230,701
Reduced Production	1,126	899	3,650	3,280,099	$	229,607
Low Production	1,078	861	1,460	1,256,942	$	87,986
		Totals	**8,760**	**7,832,766**	**$**	**548,294**
*based upon an average electrical rate		$ 0.0700	/kWh			

Proposed Arrangement	bhp	kW	Hours	KWH	Costs	
Peak Production	802	640	3,650	2,336,238	$	163,537
Reduced Production	801	639	3,650	2,333,671	$	163,357
Low Production	750	599	1,460	874,186	$	61,193
		Totals	**8,760**	**$ 5,544,095**	**$**	**388,087**
*based upon an average electrical rate		$ 0.0700	/kWh			
		Projected Savings:		**2,288,671**	**$**	**160,207**

Plastics Manufacturing Company
Mobile, AL

Vacuum Systems Financials Summary			
in US $ dollars			
Constituent	**Existing**	**Proposed**	**Variance**
1. Electricity	$548,294	$388,087	$160,207
2. Internal Labor & Overhead	$87,727	$62,094	$25,633
3. Contract Maint. & Repair	$0	$0	$0
4. Water and Treatment	$0	$0	$0
5. Scrap Reduction	$0	$0	$0
6. Rental Vacuum Pumps	$0	$0	$0
7. Other Charges	$0	$0	$0
Totals	**$636,021**	**$450,180**	**$185,840**
Estimated retrofit costs	**$125,623**		
Estimated simple payback	**0.7**	**years**	

Plastics Manufacturing Company
Mobile, AL

Item	Description	Capital	Installation	Shipping
	Prioritized Costed Action Plan			
4	Purchase and install (25) Solberg CSL-377P-800F filters for point-of-use vacuum vacuum filtration. Filters have 8" flange connection and less than 1" H2O pressure differential at full application flow.	$34,873	*	$250
5	Install automatic shutoff valves on (25) production machines. Shutoff valves should be 8" diameter and be installed in the PVC drop legs to production machines at the point of use filter. An 8" manual valve should also be installed.	$16,500	*	$500
6	Modify production machinery by installing 3" diameter tubing for vacuum supply. Vacuum supply tree should be 8" diameter PVC and use 3" full port valves for isolation. Vacuum will be drawn through (2) 3" diameter flexible tubes to each box. This modification will reduce pressure differential to machine.	$22,000	*	$500
7	Service vacuum pumps so that all controls are working properly and allowing the vacuum pump to provide full capacity to the site supply system. Oil leaks will also have to be repaird on several machines.	$10,000	*	*
8	Provide installation of the above equipment that includes mechanical and electrical work. All work is estimated and is based on performing work during normal business hours.	*	$23,000	*
10	Audit cost, project management cost and measurement/verification cost.	$18,000	*	*
	Sub Total	**$101,373**	**$23,000**	**$1,250**
	Grand Total			**$125,623**

Plastics Manufacturing Company
Existing Central Vacuum Supply Area

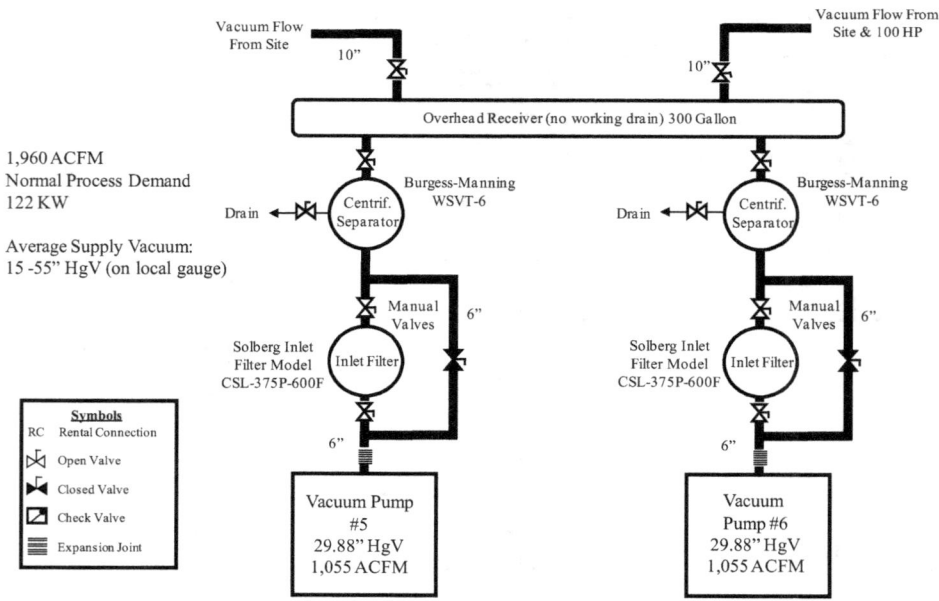

222

INDEX

NOTES:

NOTES:

NOTES: